The Open University

S216
Science: a Level Two Course

GW00722715

Environmental Science
TOPICS 9 to 11

Block 7

Cover image Forest in the foothills of Mount Taranaki (Mount Egmont), North Island, New Zealand.

The Open University, Walton Hall, Milton Keynes, MK7 6AA

First published 2002. Reprinted 2006.

Copyright © 2002 The Open University

Edited, designed and typeset by The Open University.

Printed in the United Kingdom at the University Press, Cambridge.

ISBN 0 7492 6993 6

This publication forms part of an Open University course, S216 *Environmental Science*. Details of this and other Open University courses can be obtained from the Course Information and Advice Centre, PO Box 724, The Open University, Milton Keynes MK7 6ZS, United Kingdom: tel. +44 (0)1908 653231, e-mail ces-gen@open.ac.uk

Alternatively, you may visit the Open University website at http://www.open.ac.uk where you can learn more about the wide range of courses and packs offered at all levels by The Open University.

To purchase this publication or other components of Open University courses, contact Open University Worldwide Ltd, The Open University, Walton Hall, Milton Keynes MK7 6AA, United Kingdom: tel. +44 (0)1908 858785; fax +44 (0)1908 858787; e-mail ouwenq@open.ac.uk; website http://www.ouw.co.uk

2.1

s216 block 7i2.1

GRASSLANDS

Hilary Denny and Joanna Treweek

1	Grasslands: what are they and where do they occur?	5
1.1	The importance of grasslands	5
1.2	The global grassland resource	7
1.3	Influence of climate on grassland distribution	10
1.4	Fire, grazing and the effects of human intervention on grassland distribution	12
1.5	Global variation in grasslands: some examples	14
1.6	Summary of Section 1	17
2	British grasslands: spatial and temporal variations	18
2.1	History of British grasslands	18
2.2	The variety of British grassland communities and the NVC system	20
2.3	The dynamic nature of species composition and type	27
2.4	Summary of Section 2	34
3	Grasslands: ecosystems in crisis?	35
3.1	Typical structure and dynamics of grassland ecosystems	35
3.2	Pressures on grassland ecosystems	42
3.3	Effects of increased pressure on grassland ecosystems	43
3.4	Summary of Section 3	48
4	Grassland conservation management and restoration	49
4.1	Recognition of the need for remedial action	49
4.2	Conservation of grasslands in Britain	50
4.3	Management challenges	52
4.4	Calcareous grasslands	53
4.5	Coastal and floodplain grasslands	58
4.6	Summary of Section 4	63
	Learning outcomes for Topic 9	64
	Answers to questions	65
	Acknowledgements	68

Grasslands: what are they and where do they occur?

1

Grasslands are areas of vegetation characterized by predominantly herbaceous plants (Figure 1.1). The most dominant plant groups are of course the grasses (Gramineae), sedges (Cyperaceae) and rushes (Juncaceae), but most grasslands include a significant proportion of other herbaceous plants (**forbs**) and, in some cases, there may also be scattered trees and shrubs. Grassland occurs naturally in parts of the world with intermediate rainfall, usually where there is a marked seasonal drought. It may also occur in other areas where grazing by herbivores or burning occurs frequently enough to prevent widespread establishment of woody species.

Figure 1.1 Grassland in New Zealand.

Grasslands have evolved as a distinct biome in parts of the world where net rainfall is insufficient to sustain the development of woodlands, but is in excess of that associated with desert conditions.

○ Recall from Block 3, Part 2, the six major biome types, and the two principal environmental determinants of biome type.

● The six types of biome are desert, grassland, deciduous forest, tropical forest, coniferous forest and tundra. The two principal environmental determinants are precipitation and temperature.

In the absence of direct human intervention or management, climate is the main factor that determines the presence and type of grassland. However humans, through their activities over the millennia, have had an enormous influence on the distribution and characteristics of grasslands.

1.1 The importance of grasslands

Historically grasslands have been central to the human food supply. Nearly all the major cereal crops, including wheat, rice, rye, barley, sorghum and millet, have been developed by our ancestors from wild grassland species. Many areas of grassland have been replaced by **agroecosystems**: ecosystems that are heavily managed by humans to provide food and materials from farming, either arable or pastoral. However, we still rely on grasslands to provide genetic resources (in the form of genes for disease resistance, etc.) for improving food crops, and they are still a potential source of pharmaceuticals and industrial products. Humans also see grasslands as a source of a wide range of goods and services, including water regulation and purification. Grasslands affect water supply and quality through infiltration, purification, flood control and soil stabilization.

○ Name five types of economic activity associated with grasslands.

● You can probably think of quite a few different activities but some examples would be: raising livestock such as cattle for milk, leather and meat, horse racing, pony trekking, tourism such as walking, bird watching and botanical holidays, hunting (mounted and on foot), growing grass for turf, football and other sporting events.

However, while grasslands are important economically, they also provide an important wildlife habitat. Complex and dynamic assemblages of species can develop in association with grassland vegetation, constituting a major wildlife resource.

Many grasslands provide habitats that support high levels of biodiversity, whether of the large grazing mammals of the African savannas or the diverse plant species found in traditionally managed hay meadows in northern Europe. Some areas of grassland are important for endemic species found nowhere else, for example the giant anteater (*Myrmecophaga tridactyla*) of the Argentinian pampas (Figure 1.2).

Figure 1.2   Giant anteater (*Myrmecophaga* sp.) carrying its baby on its back.

It is difficult to generalize about the importance of grassland **biodiversity** (see Box 1.1) because grasslands are such a variable habitat. However, globally, as habitat for biologically important flora and fauna, grasslands make up 19% of the Centres of Plant Diversity and 29% of ecological regions considered to be of outstanding biological distinctness.

Box 1.1 Biodiversity

The term *biodiversity* has been a bit of a buzzword since the early 1990s (see Section 4.1), and the general view seems to be that we need more of it. But what does the term mean, and is biodiversity always a good thing?

The word 'biodiversity' suggests lots of different kinds of things biological. However, does it refer to variety at the level of the ecosystem or habitat? Does it mean that within an ecosystem there are lots of different species, or that within a species or population there is a lot of genetic variation? In fact biodiversity is an imprecise term, and could be taken to mean any of these things.

As to whether biodiversity is always a good thing; in a world where fragile ecosystems such as tropical rainforests and coastal wetlands are being destroyed at a great rate, along with the plethora of animals and plants that live within them, then yes, it probably is a good thing for a number of reasons, and habitat destruction should be managed more sustainably. At the level of the individual habitat, agricultural intensification and pollution can dramatically reduce the number of species that live there. You know that the disappearance of 'indicator' species with particularly exacting requirements can give warning of a general decline in the quality of the environment. At the level of genetic variation within a species, loss of diversity can leave a species vulnerable to extinction if it cannot adapt when the environment suddenly changes, such as might occur with the advent of a particularly nasty disease (e.g. Dutch elm disease).

But, and this is an important but, individuals within a population are often particularly well adapted to that location. The term **ecotype** is used to describe a locally adapted population. So if there is an influx of individuals from a population of the same species but from a very different location, the well-adapted

combinations of genes may be broken up in the offspring, with a general loss of fitness. This loss of fitness can upset the balance of competition between species, and may even lead to the eradication of a species from a particular area.

For example, commercially available mixtures of wild flower seeds in the UK contain species that are not native to the area where they are sown. They may come from populations that grow under very different conditions, such as coastal Mediterranean populations, where temperatures are higher and the growing season starts earlier (Figure 1.3). Cross-breeding of native plants with these immigrants could result in plants starting to flower earlier, possibly before the last of the frosts, or before the emergence of the local populations of pollinating insects. The whole community could be thrown into disarray.

Figure 1.3   Is this the sort of ecotype you want in your neighbourhood?

1.2   The global grassland resource

Natural grasslands are those that form without human intervention. Extensive natural grasslands occur where the climate is generally hot (at least during the summer) and dry, although not to the extremes that result in the formation of deserts (Figure 1.4).

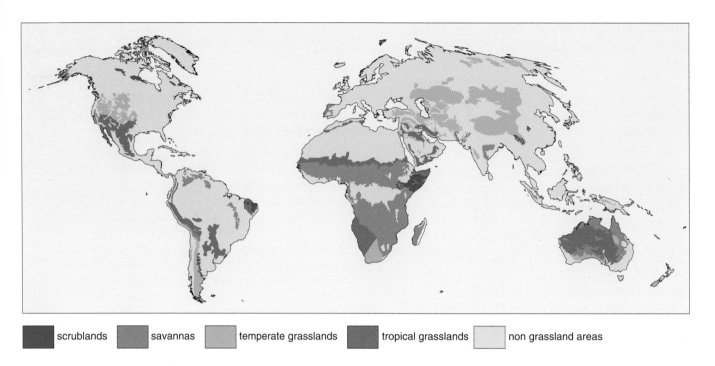

scrublands   savannas   temperate grasslands   tropical grasslands   non grassland areas

Figure 1.4   The global extent of grasslands.

Before the emergence of humans, natural grasslands occupied large areas of the continental landmasses, effectively occurring in areas of intermediate rainfall. On the basis of climate, natural grasslands are generally classified into two main types: tropical and temperate.

Tropical grasslands occur predominantly at latitudes south of the Sahara (a region known as the **Sahel**), and north of the Tropic of Capricorn in southern and eastern Africa, and in Australia. **Savanna** (Figure 1.5) forms as an intermediate vegetation type between woodland and tropical grassland, and is characterized by scattered trees or shrubs with a grass understorey.

Figure 1.5   Savanna, Tarangire National Park, Tanzania, showing the typical landscape of grassland and small scattered trees.

Temperate grasslands are found mainly in North America, Argentina, South Africa and in a wide belt from the Ukraine to China, where they are known by many different local names: **prairie**, **pampas**, **veld** and **steppe** (Figure 1.6), respectively.

Most of these temperate grassland areas have been highly exploited and substantially altered through agriculture.

○ Why do you think areas that are naturally temperate grassland, such as the North American prairie, the Russian steppe, and parts of Argentina and Australia, have become the main centres of world wheat production?

● Wheat is itself a species of temperate grass, so it performs well in those climates that support this biome.

Figure 1.6 Steppe grasslands, with nomad camp, north of Harhorin, Mongolia.

Grasslands are the potential natural vegetation on $33 \times 10^6 \, \mathrm{km^2}$ (approximately 25%) of the land surface of the Earth. However the world's actual area of grassland is closer to almost twice that, due to the existence of large areas that are to some extent artificially maintained as grassland through human intervention (see Section 1.4). This figure is remarkably close to that estimated for grassland extent in the year 1700 AD, suggesting that the factors affecting the global area of grassland have maintained a rough balance over the past 300 years. In fact this apparent stability represents a dynamic equilibrium, because, globally, it has been maintained by losses through conversion of grasslands to cropland being balanced by gains through deforestation. However, trends vary regionally, as the data in Table 1.1 show.

Table 1.1 Estimated percentage of natural grassland remaining and converted to other uses in selected continents.

Continent and grassland type	Remaining as grasslands	Converted to croplands	Converted to urban areas	Total converted
N. America: tall-grass prairie	9.4	71.2	18.7	89.9
S. America: pampas, etc.	21.0	71.0	5.0	76.0
Asia: steppe, etc.	71.7	19.9	1.5	21.4
Africa: veld, miombo, etc.	73.3	19.1	0.4	19.5

○ Using the data in Table 1.1, how do the figures for North America and Asia compare?

● In North America nearly 90% of natural grasslands have been converted to other use, compared with around just 20% in Asia. Most of this conversion is to agricultural land in both continents, but in North America nearly 20% of the conversion is to urban area, compared with just 1.5% in Asia.

The reasons for the differences in rates of conversion of grassland to croplands and urban areas (as well as forest to grassland) are largely tied to increases in population density, a subject that is dealt with in more detail in Section 3.2.

Question 1.1

Explain briefly what is meant by the term 'grassland'.

Question 1.2

Roughly what proportion of the Earth's land surface is occupied by grassland? List the principal reasons why grasslands are seen as important.

1.3 Influence of climate on grassland distribution

In general, tropical grasslands receive 500–1500 mm of rain in an average year and experience temperatures of around 15–35 °C, with a dry season possibly lasting as long as eight months. Figure 1.7 shows a climate diagram for an area of savanna near Harare in Zimbabwe. These diagrams are similar to the plots of monthly rainfall and potential evapotranspiration that you met in Block 3, except that the two lines represent mean monthly precipitation and mean monthly temperature. A biogeographer, named Heinrich Walter, showed that the monthly potential evapotranspiration (in mm) could be roughly equated to the mean monthly temperature (in °C) multiplied by two. This approach is widely used in practice, as it removes the difficulty of calculating evapotranspiration rates exactly. The diagrams, when drawn in the format proposed by Walter, allow the climate of an area to be assessed quickly. Times of year when the line denoting temperature is above the line denoting rainfall are regarded as periods of relative drought.

Figure 1.7 Walter climate diagram for an area of savanna near Harare, Zimbabwe.

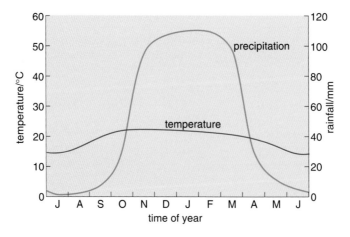

○ When and roughly how long is the period of relative drought in Zimbabwe?

● Zimbabwe experiences drought from early April to late October, a period of seven months.

Temperate grasslands experience large seasonal changes in temperature, as much as 40 °C. Mean annual rainfall in North American grassland areas is 300–600 mm, with winter temperatures ranging from −18 °C in the north to 10 °C in the south, with corresponding summer values of 18 °C in the north and 28 °C in the south.

Figure 1.8 shows a climate diagram for steppe grassland around Kabul in Afghanistan.

○ When and roughly how long is the period of relative drought in Afghanistan, and when are plants likely to experience sub-zero temperatures?

● Afghanistan experiences drought from early May to mid-November, a period of nearly seven months. Sub-zero temperatures can occur any time between early October and late April: seven months of the year.

In both tropical and temperate situations, natural grasslands occur where growing conditions are optimal for a limited period within the year. In tropical grasslands, the main growth period is usually restricted to the rainy season. In temperate grasslands, growth tends to occur in a short period between the cold, damp winter and the hot, dry summer.

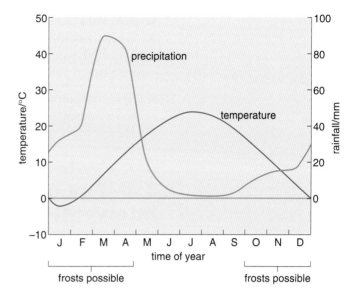

Figure 1.8   Walter climate diagram for steppe grassland near Kabul in Afghanistan.

If plants are to exploit these conditions successfully they must have stored reserves of energy to enable rapid shoot production, flowering and seed-setting, all within the limited growth period. Perennial grasses have evolved the ability to complete their reproductive cycle rapidly during a limited growing season. Their subterranean perennating roots and rhizomes enable them to survive periods of drought or cold, and also to recover from the effects of fire (see Section 1.4).

Latitude and altitude

Some typical types of grassland are found at high latitudes or altitudes (above the mountain tree line), where it is too cold for trees to grow.

Tussock-forming species often characterize these areas, particularly in the Southern Hemisphere. Examples of tussock grasslands can be found as the main vegetation type of the subantarctic islands of the Southern Ocean and also in the drier, colder areas of New Zealand and the southernmost parts of South America (Figure 1.9).

Waterlogging

Seasonal flooding or waterlogging may prevent the growth of woody species, thereby favouring the predominance of grass species. This tends to occur in the subtropics where there is considerable meteorological variation between the seasons. In the Mato Grosso region of Brazil, the 140 000 km² of subtropical grassland known as the Panatal is dry for six months of the year and is then flooded during the wet season, producing extensive shallow wetlands. Small pockets of woodland occur on slightly elevated areas that do not flood.

Figure 1.9   In New Zealand, tussock grassland is common in areas too cold for tree growth.

○ Given that natural grasslands occur where the temperatures are too low, or where conditions become seasonally too dry or wet for tree growth, what do you think would happen to grasslands if the climate became warmer or the patterns of precipitation became less extreme?

● The grasslands would gradually change to woodland because, where environmental conditions permit tree growth, trees are stronger competitors for resources than are grassland species.

1.4   Fire, grazing and the effects of human intervention on grassland distribution

Fire and grazers are naturally occurring factors that lead to the removal of the shoots of grassland vegetation. Grassland plants are able to withstand this constant cropping because their growing points are often at ground level: when leaves are removed, the growing points are not. Furthermore, in the leaves of forbs and broadleaved trees, cell division and expansion occur throughout the leaf, whereas in those of many grasses and their allies these processes occur only at the base of the leaf. Consequently, when the upper portion of a leaf is cut or grazed, the grass-type leaf can continue to expand from the base to replace the portion removed. Trees in the early stages of development are especially intolerant of repeated shoot removal. So even when climatic and soil conditions are suitable for tree growth, the presence of sporadic fires or a significant number of grazers can prevent woodland development.

Over many centuries, humans too have modified the natural vegetation. Grasslands are used to provide forage for domestic livestock, housing materials, fuel and some crops for human consumption (millet, for example). Humans have also extended the area of grasslands by cutting down forests. Most human interventions involve relatively frequent harvesting of biomass (living organic matter) and inhibition of the establishment and growth of woody plants.

In the absence of other factors, seasonal drought is a key factor promoting the establishment of grasslands, but removal of biomass by herbivores, cutting or by fire can override the influence of climate and cause grasslands to occur in areas that might otherwise be occupied by woodland.

○ Recall from Block 3, Part 2, the term used to describe vegetation that is maintained by repeated disturbance.

● Vegetation maintained in this way is called a *deflected climax*.

Many grasslands are considered to be deflected climaxes. In large parts of the world, in both temperate and tropical latitudes, the production of **semi-natural grasslands** has followed deforestation. In these grasslands, prevalence of grazing or burning arrests natural succession to woodland and results in the maintenance of grassland. These grasslands may share a number of species in common with naturally occurring (*climatic climax*) grasslands, but species balance and vegetation structure are noticeably altered as a result of human intervention.

Intensification of human intervention (for example through addition of fertilizers or reseeding) results in 'intensive' or **agricultural grasslands**, which are generally characterized by a limited number of grass species that are managed as a crop. These grasslands tend to have limited value for wildlife conservation.

In many parts of the world, spontaneous fires occur once grasses have died back after flowering and there are large volumes of dry biomass. In addition to the spontaneous fires of natural grassland, fire has been used to manage grasslands, with a long-standing impact on the character of many grassland habitats (Figure 1.10).

Figure 1.10   Grassland fires are often started deliberately by humans, as part of the management regime.

Burning is used deliberately to remove dead plant material, to reduce numbers of livestock parasites and to encourage fresh, new growth for feeding domestic animals. Many tropical and subtropical grasslands are believed to be maintained by fire. While spontaneous fires caused by lightning strikes may occur, an increasing proportion of fires are human-induced. Because fire use has such a long history, it has probably had a major impact on present ecosystem structure and function in many areas. The savanna ecosystems now found in Cuba are the product of long-term human impact, that is combined cutting, burning and grazing. Recent studies of the prairies in North America interpret them in part as semi-natural ecosystems maintained by grazing of wild animals, but also as systems maintained by fires deliberately set by Native Americans. In the absence of these influences, trees and shrubs soon encroach.

Question 1.3

Allocate the grassland types listed below to the appropriate characteristics in Table 1.2.

Grasslands types: natural, semi-natural, intensive or agricultural, tropical, temperate.

Table 1.2   Characteristics of selected grassland types.

Grassland type	Characteristics
	Often highly exploited and substantially altered through agriculture; they have become the main centres of world wheat production.
	Managed deliberately to produce forage crops and usually consist of a limited number of species.
	Sometimes characterized by scattered trees or shrubs with a grass understorey; generally exist between the belts of tropical forest and desert vegetation.
	Distribution and species composition are determined by climate and soil type.
	Communities representing deflected climaxes that are maintained by some human intervention, such as deforestation followed by regular burning or grazing.

Question 1.4

Use the appropriate words from this list to complete the following passage: short, long, close to, well above, underneath, rainfall, temperature, precipitation, winds, fire, grazing, disease, fertilizers, shoot, root, hemicryptophyte, geophyte, marked, imperceptible, energy, chlorophyll.

The most common functional type (from Raunkiaer's classification, see Block 3, Part 2) to be found in grasslands is a ————. These types of plants are well suited to the grassland environment because their growing points are positioned ———— the ground. This enables them to avoid damage caused by ———— seasonality in terms of ———— and ————, as well as the damage caused by ———— or ————. Most grassland plants complete their reproductive cycle over a ———— period of time, and they must have stored reserves of ———— to enable rapid ———— production, flowering and seed-setting.

1.5 Global variation in grasslands: some examples

Figure 1.11 Short-grass prairie grassland, Colorado, USA.

Figure 1.12 The grey wolf (*Canis lupus*), a top predator of the North American prairies.

Around the world, natural and semi-natural grasslands are associated with characteristic flora and fauna. It is likely that almost all grasslands have experienced a long history of human intervention through different forms of management. These interactions will have played their part in defining the character and composition of the grasslands over time.

North America

In the North American prairies (Figure 1.11), species composition varies between regions. The central and northern areas are characterized by *Stipa* spp. and *Festuca* spp., respectively; the western region has a short steppe community dominated by *Bouteloua gracilis,* and in the east the tall-grass prairie is characterized by the bluestem grasses, *Andropogon* spp.

Before the prairies were affected by European colonizers, huge herds of bison (*Bison bison*) grazed across the prairies; together with pronghorn (*Antilocapra americana*), bison were the main herbivores of the region. Grey wolves (*Canis lupus*, Figure 1.12), which occur through much of North America, predated on the grazers and other smaller animals of the prairie, including hare and rabbit species and burrowing rodents. Grasshoppers were, and remain, the most significant invertebrate component of the invertebrate fauna.

South America

In South America, the grasslands, primarily located across the southeast of the continent, can be subdivided into the pampas of Argentina, characterized by species of *Stipa*, and the **campos** of

Uruguay and Brazil, characterized by a predominance of species of *Andropogon*. Prior to the many changes wrought by centuries of heavy grazing and burning, the principal large herbivore was the pampas deer (*Ozotoceros bezoarticus*). Other species native to these grasslands include jaguar (*Panthera onca*), armadillo and the rhea (*Rhea americana*, Figure 1.13), a large, flightless bird endemic to the area.

Eurasia

The other significant belt of temperate grassland, the Eurasian steppe, extends from the Ukraine to Mongolia and bears many similarities in species composition with the prairie of North America. Species of *Stipa* and *Festuca* are common constituents of the flora, variously assuming dominance in the different areas. Burrowing rodents, including large marmots (*Marmota caudata*, Figure 1.14) are important members of the fauna, playing key roles in the ecosystem through their feeding and burrowing habits.

Africa

The largest of the tropical grasslands is the Sahel of sub-Saharan Africa (Figure 1.15). The main grassland components today include species of *Aristida* and *Schoenefeldia,* but these tough grasses may have been less dominant in the past compared with more palatable species that are now largely grazed out. Parts of the Sahel are maintained as grassland by burning, grazing and fuel gathering, where they would naturally revert to savanna under the absence or relaxation of these practices.

Figure 1.13   The rhea (*Rhea americana*), a characteristic bird of the South American pampas.

Figure 1.15   Sub-Saharan grasslands, Waza, northern Cameroon, West Africa.

The tropical grasslands of East Africa are generally more diverse than those of the Sahel as they receive higher rainfall. Vegetation patterns vary according to the climate. In the drier sites, *Aristida* species are important. Tall grasslands of *Pennisetum* tend to predominate in the wetter regions where forest cover has

Figure 1.14   The red or long-tailed marmot (*Marmota caudata*), eastern Pamir, Tadzhikistan.

Figure 1.16   White rhinos (*Ceratotherium simum*) are characteristic herbivores of the tropical grasslands of East Africa.

been removed and the habitat is kept open by the grazing of large herbivores, such as elephants, or by burning. Among the fauna, wildebeest, antelope, rhinoceros (the white rhinoceros, *Ceratotherium simum*, is shown in Figure 1.16), buffalo and elephant are, or have been, significant herbivores, with carnivores including lions, cheetah and hyena.

Australia

Australian tropical grasslands of the drier, central regions are characterized by the spinifex grasses *Plectrachne* and *Triodia*, which typically form hummocky grassland as they trap wind-blown sand at the tussock bases. The leaves of spinifex grasses are tough and spiky, but the seed heads provide good fodder. In the damper areas in northern Australia other species such as *Sorghum* (Figure 1.17) dominate. In drier areas, the introduced tree species *Acacia nilotica* can invade the vegetation to the extent that it can no longer be classified as true grassland.

Figure 1.17   Spear grass (*Sorghum intrans*), a species with fire-resistant seeds, Australia.

Kangaroos and wallabies (Macropodidae) are the characteristic large herbivore associated with Australian grassland (Figure 1.18). Another native animal of the grasslands is the emu, a large, flightless bird. Introduction of domestic grazing stock and other non-native species has significantly changed the balance of the fauna over the years since human settlement. Cattle, sheep, rabbits, camels, horses and goats are all widespread and abundant. The activities of sheep and rabbits have played a large part in changing the native ecology of the grasslands.

Figure 1.18   Wallabies are characteristic large herbivores of Australian grasslands.

○ What do the faunas of the grassland biomes around the world have in common?

● A diversity of associated animal species, but, most notably, the presence of large herbivores such as bison, antelopes, rhinos and kangaroos.

Question 1.5

For each plant or animal listed in Table 1.3, choose the appropriate terms to describe the associated type of grassland, and its role within the community.

Grasslands: Sahel, prairie, steppe, Australian tropical grassland, pampas, E. African tropical grassland.

Roles: primary producer, herbivore, carnivore.

Table 1.3 Role of selected organisms in different types of grassland.

Organism	Type of grassland	Community role
spinifex grass		
white rhinoceros		
marmot		
jaguar		
Schoenefeldia spp.		
Stipa spp.		
N. American bison		

1.6   Summary of Section 1

1   Grasslands are areas dominated by grassy vegetation and maintained by fire, grazing and drought or freezing temperatures. They encompass non-woody grasslands (temperate and tropical), savannas and scrublands.

2   Grasslands cover almost half of the Earth's land surface and are important economically for food production and recreation, as well as being a reservoir of biodiversity. Over the last 300 years, losses of grassland through conversion to arable and urban development have largely been balanced by gains through deforestation.

3   Grasslands are divided into three main types. (i) Natural grasslands, the distribution and species composition of which are determined by climate. (ii) Semi-natural grasslands, in which some human intervention (e.g. deforestation followed by regular burning or grazing) results in grasslands that may retain many of the species found in natural grasslands, but have a noticeably different balance of species and vegetation structure. (iii) Intensive or agricultural grasslands, which are managed deliberately to produce forage crops, and usually consist of a limited number of species.

4   The main determinants of grassland distribution are climate (mainly temperature and precipitation), burning and grazing, the last two of which are heavily influenced by human activity.

5   There are many different types of natural and semi-natural grassland community worldwide. They differ greatly at the species level, but functionally they have many similarities. The vegetation is dominated by hemicryptophytes that can withstand removal of shoots through grazing or fire, as well as marked seasonality (i.e. periods of relative drought, waterlogging or cold). The fauna is characterized, most noticeably, by the presence of large herbivores.

2

British grasslands: spatial and temporal variations

Britain was once a predominantly forested group of islands. Today it consists of a complex mosaic of vegetation types that have altered greatly throughout history (Figure 2.1). Patterns of vegetation, whether British or otherwise, are dynamic in that they vary in both space and time. The extensive tracts of grassland that have developed over the years had their origins in the woodland clearances carried out by people, possibly thousands of years ago, in order to create settlements and engage in agricultural activities.

Figure 2.1   A typical mosaic of vegetation types in the west of Britain.

Forms of management carried out over hundreds of years have resulted in the formation of many specialized assemblages of plants and animals. These semi-natural grasslands can be very complex and difficult to recreate once damaged or destroyed. In countries with a long history of human intervention, like Britain, where little 'natural' wildlife habitat remains, such semi-natural grasslands are often an important wildlife resource.

You now know that, on a global scale, the distribution of natural grasslands is largely determined by climate, but that the extent and type of grassland is heavily influenced by human activity. At a more local scale, the composition and balance of species in a grassland is often influenced by a number of other biotic and abiotic factors.

Research work carried out by botanists in the latter half of the 20th century has given us a framework for describing and classifying the many different types of grassland community, as well as a better understanding of the functional relationships between them. This understanding should help us to devise more sustainable ways of conserving and managing our grasslands.

2.1   History of British grasslands

British grasslands have had a chequered history. Archaeological evidence from various parts of Britain shows that during the Bronze Age (2600–3600 BC) there was a move away from the arable agriculture that had developed during the Stone Age, and towards grazing systems. It is believed that this was in response to the climate becoming cooler and wetter, leading to more frequent crop failures in the old arable system. This shift in agricultural management coincided with large-scale forest clearance and the expansion of grasslands.

○   Why should switching reliance from crops to livestock as a food source create the need for additional agricultural land?

●   The human population became predominantly secondary consumers rather than primary ones. Being at a level higher in the food chain is less efficient and requires more primary production to sustain the same population.

Traditionally, grasslands were managed around the needs of farm animals. Most farms kept animals both as a source of food and to pull the ploughs, feeding them on **pasture** for much of the year (Figure 2.2a). However, during the winter and early spring, when the grass is not growing, dried grass stored as hay is needed as an alternative food source. Hence on most farms, the best areas of grass growth were reserved as hay **meadows** to sustain livestock throughout these lean months (Figure 2.2b).

(a)   (b)

Figure 2.2   (a) Livestock grazing on pasture. (b) Modern hay bales in a meadow.

At the time of the Domesday Book in 1086, it would appear that most of the large downlands (i.e. the treeless, undulating chalk uplands of southern England) were already in existence and used for grazing sheep (Figure 2.3), whilst meadows were numerous but not extensive.

Figure 2.3   Image from illuminated manuscript showing late Saxon shepherds with their flocks.

By the Middle Ages the population had increased; large areas of land were under the plough, and grassland had become scarcer but more intensively exploited. More land was managed as meadow and it could command a value four times greater than that of arable land. Some meadows were privately owned, whereas others were held in common.

The Black Death of 1349 drastically reduced the population and much arable land was consequently converted to grassland, due to lack of labour to maintain land in cultivation. Later in the Middle Ages, a boom in the wool and cloth trade increased the need for sheep pasture. However, from the middle of the 18th century, the Agricultural Revolution started to introduce great changes to land management, as areas of ancient grassland were brought under cultivation. Thereafter followed the Enclosure Acts, which abolished much common land and led to the conversion of more grassland and heath to arable or agricultural swards. Small areas of common land reverted to woodland.

○ What seem to be the dominant type of factors affecting the character of the semi-natural grasslands of Britain through history?

● The principal factors instigating changes in the character of British semi-natural grasslands appear to be mainly socioeconomic.

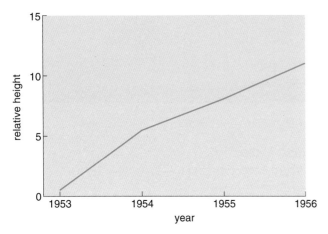

Figure 2.4 Relative vegetation height on an area of downland from the year before the 1954 myxomatosis outbreak until two years afterwards.

More recently the countryside has again changed dramatically. In 1953, myxomatosis, a viral disease, devastated the rabbit population. Figure 2.4 shows what happened to the relative height of grass vegetation in the two years that followed the outbreak.

○ What is the effect of the loss of the herbivorous rabbits on the height of the vegetation on the down?

● The vegetation shows a linear increase in height. Two years after the epidemic the vegetation is about ten times higher than before the outbreak.

Once the level of herbivory fell, coarser vegetation gained dominance, tree seedlings became established more easily, and communities changed, often to the detriment of rare and specialized grassland species. The effect on short turf swards and heathlands was devastating.

2.2 The variety of British grassland communities and the NVC system

Except for scattered fragments in remote and inaccessible places, such as on some mountain or cliff ledges where they have escaped grazing or other forms of disturbance, there are no truly natural grasslands left in Britain. Existing swards are either agricultural or at best semi-natural.

○ Recall the two main types of managed grassland that predominate in rural Britain.

● The two types of managed grassland are meadow that is mown for hay, and pasture that is grazed by livestock.

The term 'semi-natural' does not imply artificial in the sense of having been deliberately sown. Indeed, the communities of native grasses and forbs evolved naturally in response to regular, consistent management over many years, such as winter grazing, summer cutting or seasonal flooding. Withdrawal of management results in succession to woodland within a few decades.

Swards resulting from artificially sown seed mixtures, such as sports grounds, parks and road verges, are common in urban areas and are known as amenity grasslands. The use of seed mixtures is also becoming more common in agricultural settings, wherever farmers want 'instant' pasture. Such swards tend to lack diversity.

Mowing is indiscriminate in the plants it affects, whereas livestock may graze some plants preferentially. These factors, along with the timing of management, influence the community composition of grassland. Interacting with management practices at a local level are a variety of abiotic variables that are superimposed over the major climate variables of temperature and precipitation.

○ Recall from the Block 3, Part 2, the key abiotic variables that are likely to influence the general composition of semi-natural grassland.

● The principal abiotic variables influencing grassland composition are the underlying geology and resultant soil type, aspect, microclimate, topography and hydrology.

The result of all this local variation is that the vegetation of the British Isles is characterized by a diverse range of grassland types. The National Vegetation Classification (NVC) divides grasslands into three categories.

○ Recall from Block 3, Part 2, the major categories for grasslands used by the NVC system.

● The NVC divides grasslands into mesotrophic; calcicolous; and calcifugous and montane communities.

However, many different types of community are included within these three major types. Within mesotrophic grasslands, 13 distinct communities are recognized; 14 are described under calcicolous grasslands; and for calcifugous grasslands and montane communities, 21 communities are identified. Within each of these communities one or more subcommunities may be recognized. Use of this system enables an accurate profile of a plant community to be produced. This can be extremely valuable when planning management or considering the impact of some form of disturbance upon the community.

Factors determining the distribution of the major grassland types

Mesotrophic grassland is found where the pH is between 5 and 6.5; calcareous grassland is generally found on shallow soils with a pH range of 6.5–8.5; calcifugous and montane grassland normally occurs on acid soils with a pH lower than 5.

○ Recall from Block 3 what the principal environmental variables are that determine soil pH.

● Underlying rock type and precipitation.

Mesotrophic grasslands are the dominant grassland type in England and they are scattered throughout the British Isles (Figure 2.5). During the period of research that led to the establishment of the NVC system, representative samples were taken from all the key areas of vegetation in the British Isles (excluding Ireland).

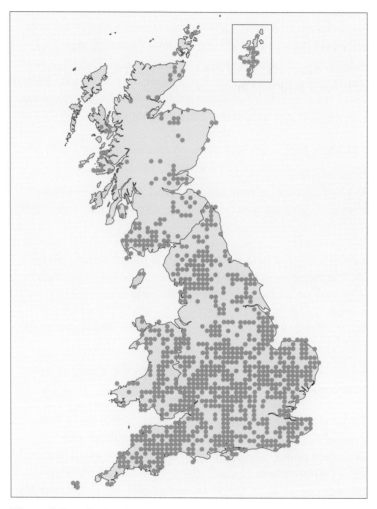

Figure 2.5 Distribution of samples originating from mesotrophic grasslands (taken during NVC research work).

Mesotrophic grasslands, as the name suggests, are the ones that grow on soils with moderate nutrient availability, which favours the growth of the majority of plant species. The soils tend to be deep, often with a high clay content. Mesotrophic grasslands occur on soils that are the most favourable for the growth of arable crops and lush pasture, suitable for cattle raising.

The distribution of calcareous grassland is shown in Figure 2.6. Typically, calcareous grasslands occur on the chalk and limestone country of the Chilterns and the Cotswolds (Figure 2.7). They occur on soils that are often quite shallow and infertile.

○ Recall from Block 3, Part 2, the term used to describe the 'lime-loving' plants that occur in calcareous grassland.

● The term used to describe lime-loving plants is calcicoles.

○ Many calcareous rocks are rich in nutrients, so why are soils derived from them so infertile? Refer back to Block 3 if you cannot remember.

● Calcareous soils have a high pH, and some key plant nutrients such as phosphates are relatively insoluble under these conditions.

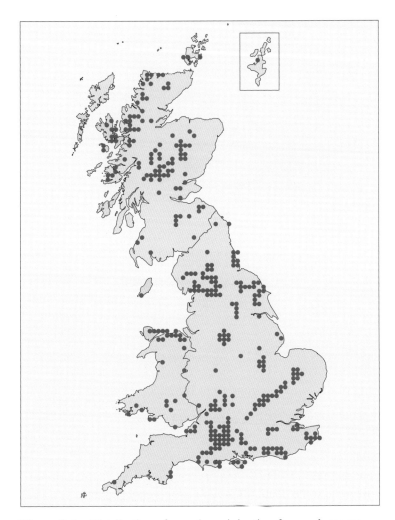

Figure 2.6 Distribution of samples originating from calcareous grasslands.

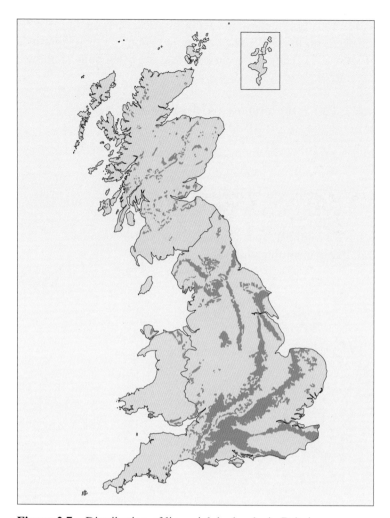

Figure 2.7   Distribution of lime-rich bedrocks in Britain.

Typically, calcicolous plants are slow growing, and consequently the vegetation tends to be short. Before the advent of modern fertilizers, this type of grassland was used mainly as pasture for sheep, which are better able to nibble the short turf than are either cows or horses.

The distribution of calcifugous and montane vegetation is shown in Figure 2.8

○   Recall the meaning of the term calcifuge from Block 3.

●   The term calcifuge means 'lime-hating'.

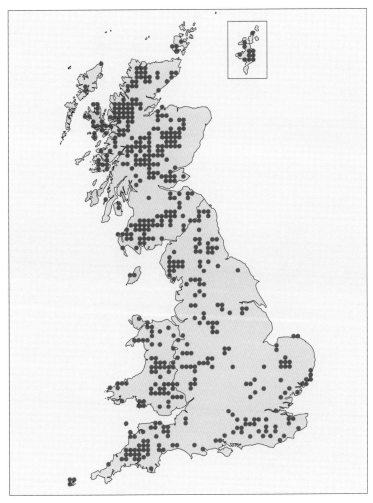

Figure 2.8 Distribution of samples from calcifugous and montane vegetation.

Calcifuge and montane vegetation occur on acidic, nutrient-poor soils.

○ What do you think are the main reasons responsible for the lack of fertility in soils underlying this type of vegetation?

● These soils become infertile either because the underlying rocks are low in nutrients and the pH is so low that several key nutrients are insoluble, such as occurs over slate, or because conditions promote the removal of soil nutrients by leaching, as on sandy soils or on steeply sloping ground in areas of high rainfall (Figure 2.9).

Figure 2.9   The main areas of Britain over 200 m, with the highest mountains shown in dark green. In Britain, the west and mountainous areas also receive the highest rainfall.

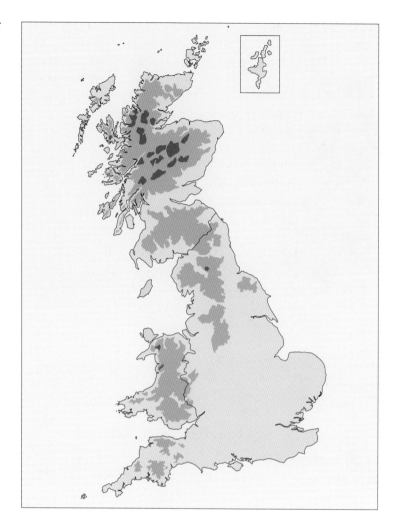

The principal uses of these acid grasslands are as sheep pasture in the north and west, although many acid grasslands of this sort are utilized as military training areas.

○   What then appear to be the major factors determining the soil pH and therefore the distribution of these three major types of grassland in Britain?

●   The major factors are: climate, especially precipitation, underlying geology (and therefore the associated soil type) and topography.

Question 2.1

Select words from the list below to complete the passage that follows, about the history of British grasslands.

water meadows, cattle, 18th, meadowland, Corn Law, Dark Ages, pastoral, Tudor period, arable, meadow, downlands, pasture, seven, three, four, wool and cloth trade, Black Death, foot and mouth disease, less, cholera, more, beef and leather trade, sheep, 17th, 19th, Enclosure, common land, arable land, myxomatosis, Middle Ages.

At the time of the Domesday Book in 1086, it would appear that most of the large ———— were already in existence, whilst ———— was common but not extensive. By the ————, large areas of grassland had been converted to arable use. The remaining grassland was ———— intensively utilized. More land was managed as ————, which had a value ———— times higher than arable land. During the 1300s, the ———— drastically reduced the population; much ———— land was converted to grassland. Later in the Middle Ages, a boom in the ———— increased the need for ———— pasture. However, in the ———— century, the Agricultural Revolution wrought great changes to the land. Thereafter followed the ———— Acts, abolishing much ————. In 1954, an outbreak of ———— had a severe effect on the short turf swards.

Question 2.2

Some of the criteria used to classify different grassland types are listed in Table 2.1. Recall the grassland encountered in Activity 2.1 of Block 3, Part 2. Taking the main community type at Hollington Basin, the one dominated by tufted hair grass (*Deschampsia cespitosa*), categorize it according to each of the criteria listed in Table 2.1.

Table 2.1 Various criteria used to classify grassland types.

Criteria used to classify grassland type	Categories
climate	temperate, tropical
soil pH	calcifugous, mesotrophic, calcareous
soil moisture	wet, dry
degree of modification by humans	natural, semi-natural, agricultural
primary management	meadow, pasture, sports turf
botanical composition	NVC types, e.g. CG8, MG6

2.3 The dynamic nature of species composition and type

At a local level of spatial and temporal scale, climate, underlying geology and topography are fairly immutable variables. They certainly change very slowly when looked at from the human perspective. So what factors determine the distribution of the subsidiary communities of grassland encompassed by the three major types? What are the predominant variables controlling the precise composition of a grassland community at the local level?

○ What do you think are the key factors affecting community composition at a local level?

● Soil type, hydrology, topography, aspect, microclimate and land management are probably the most important, but other factors that can be important are proximity to other habitats, size of the grassland, and biotic factors such as the impact of invertebrate and herbivore populations, and the presence or absence of predators.

Indeed at this local level, biotic factors, especially those that are anthropogenic (result from human activity) can become very important.

It should have become apparent to you earlier in the course that individual species respond to environmental factors in characteristic ways. It is often possible to recognize 'distribution gradients' for species in relation to environmental factors. The percentage constancy of occurrence of selected grass species in relation to one variable — soil pH — is shown in Figure 2.10. (Refer back Block 3, Part 2 if you cannot remember what is meant by a 'constant' species in a community.)

Figure 2.10   Distribution gradients for selected grass species along a pH gradient in England: (a) mat-grass, (b) quaking grass.

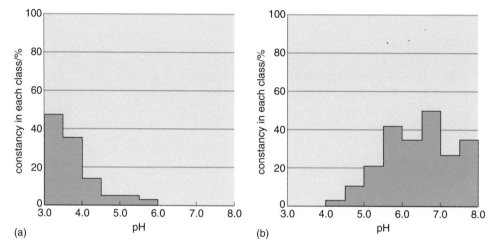

○   What is the relationship between pH and the occurrence of mat-grass and quaking grass.

●   Mat-grass occurs under acidic conditions (pH < 6), whereas quaking grass (*Briza media*) occurs at neutral and high pH.

Therefore when the differential tolerances of each species to all the relevant variables are taken into consideration, it can be seen that even quite small changes in one or more environmental variable can lead to a change in community composition.

○   Recall the term used to describe the 'hypothetical space' occupied by each species when its unique ranges of tolerance to all variables are taken into account.

●   The unique hypothetical space occupied by a single species is known as its niche.

The NVC classification system recognizes a discrete number of communities and subcommunities within each major grassland type. However, in the field grasslands tend to grade into one another gradually, forming ecotones rather than clearly demarcated areas. Nevertheless, grasslands can be broadly characterized according to the environmental conditions or circumstances in which they develop and the management imposed on them, as well as on the basis of botanical composition. A considerable range of grassland types occurs, with clear distinctions between them, as the following examples show.

Soil type

The relationships between soil nutrients and diversity are complex and not fully understood. Research results for British grasslands suggest that for high diversity and species richness to occur, the amount of phosphorous extractable from the soil should be less than 20 mg kg^{-1}. In British soils, the evidence suggests that phosphorus is often the limiting nutrient.

There is a clear range of different grassland communities along a gradient of soil type, the principal variables being organic matter content and moisture status. Figure 2.11 illustrates the range of calcifugous grassland communities along this gradient.

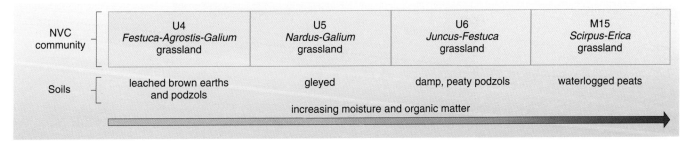

Figure 2.11 Sequence of calcifugous grasslands and wet heath in relation to soil variation.

Festuca spp. and *Agrostis* spp. are typical fine-leaved pasture grasses. *Juncus* spp. are rushes, and indicate noticeably wet soils (Figure 2.12), while *Erica tetralix* is cross-leaved heath, one of the heather family that prefers wetter conditions. Mat-grass (*Nardus stricta*) is a very tough grass, characteristic of heavily grazed upland pastures (Figure 2.13). In fact it is now realized that *Nardus stricta*-dominated grassland often replaces heather moorland when grazing pressures are raised above a certain threshold.

Figure 2.12 *Juncus* spp. in an area of wet grassland.

Figure 2.13   Mat-grass (*Nardus stricta*) on heavily grazed upland pasture. Dark heather growth is visible only where sheep are excluded.

Many heather moorlands have declined in recent years, with a consequent loss of aesthetic value as well as threatening the populations of moorland birds such as red grouse (*Lagopus lagopus scoticus*, Figure 2.14).

○   What do you think would be the most effective step one could take to restore the heather moorland in such circumstances?

●   Reducing the stocking levels of sheep would probably be the most effective change in management in order to restore heather moorland.

Figure 2.14   Red grouse (*Lagopus lagopus scoticus*) numbers have fallen as heather moorland has declined.

Sites where the water-table remains close to the surface over most of the year can give rise to **wet grasslands** and **fen**. The pH of these communities can range from acid (pH 4) to base-rich (pH 7.5). Wet grassland develops on land that is periodically flooded or waterlogged by freshwater. Wet grasslands include semi-natural floodplain grassland, water meadows, and wet grassland on poorly draining soils. Similar areas, if left ungrazed, develop into fen, which is typified by taller vegetation such as the common reed (*Phragmites australis*); this is a member of the grass family but often grows to more than two metres in height.

Figure 2.15   Alder (*Alnus glutinosa*) and willows (*Salix* spp.) invading fen vegetation.

Fen vegetation (Figure 2.15) evolves naturally into woodland. Species such as sedges (*Carex* spp.), rushes (*Juncus* spp.), birch (*Betula* spp.), alder (*Alnus glutinosa*) and willows (*Salix* spp.) invade the dense reedbeds that develop around open water. This succession occurs because decomposition is slow in the waterlogged conditions, and the gradual accumulation of dead plant material slowly raises the land surface above the water-table.

○ How are wet grassland habitats stabilized? That is to say, why do they not all become woodland over time?

● Cutting and/or grazing prevents fen and woodland species colonizing the grassland.

Management

Figure 2.16 illustrates the range of mesotrophic grasslands that occur along a gradient of differential cropping and fertilization.

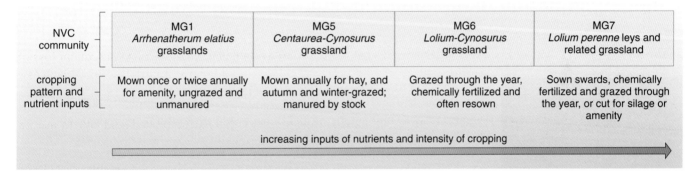

Figure 2.16   Mesotrophic pastures and meadows in relation to levels of cropping (by grazers and mowing) and fertilization (by manuring and application of chemical fertilizers).

○ Using Figure 2.16, which species of grass appears to flourish in conditions of high fertility and cropping, and which in conditions of relatively low fertility and low cropping?

● *Lolium perenne* flourishes in high levels of fertility and cropping, while *Arrhenatherum elatius* can thrive in conditions of lower fertility in the absence of regular cropping.

Lolium perenne (perennial rye-grass) has the characteristic of being very resistant to wear and tear. Consequently, it is a major constituent of many commercial grass seed mixtures, and as such is widely sown in amenity and agricultural grassland contexts. The role of management practices is discussed in more detail in Section 4.

Microhabitats

It should now be clear that quite minor changes in conditions, over small distances, can bring about shifts in the floristic composition of a community. A range of NVC communities may be present as a result of subtle differences in microhabitat — a situation exemplified by the variety of grassland communities that typically occur in and around an irregularly grazed lowland pasture (Figure 2.17).

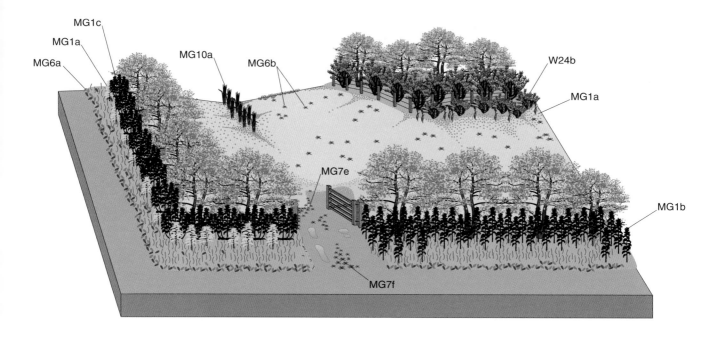

Figure 2.17 Typical pattern of grasslands in and around a run-down lowland pasture.

The most characteristic community of irregularly grazed mesotrophic grassland throughout Britain is the *Arrhenatheretum* MG1. This community occurs on road verges, railway embankments and in neglected agricultural and industrial habitats, such as badly-managed pasture and meadow, and building sites. The *Arrhenatherum elatius* community is dominated by coarse-leaved tussock grasses, such as the false oat (*Arrhenatherum elatius*), cocksfoot (*Dactylis glomerata*) and Yorkshire fog (*Holcus lanatus*). Large umbellifers such as cow parsley (*Anthriscus sylvestris*, Figure 2.18a) and hogweed are frequent, as are thistles and nettles (*Urtica dioica*). Below this taller vegetation is a layer of fine-leaved grasses such as rough meadow-grass (*Poa trivialis*), and small forbs such as dandelion (*Taraxacum* spp., Figure 2.18b) and dock (*Rumex* spp.). There are five subcommunities recognized, which thrive on well-structured and freely draining loams. The most important soil variables influencing floristic variation are pH, nutrient availability (especially N, P, K), and drainage.

Festuca rubra is a common component of the community MG1a, where calcareous deposits may raise pH. Nutrients tend to be leached out of soils, and then accumulate around drainage ditches. Even the tendency for birds to perch (and so defaecate) on walls and gates can lead to local accumulations of nutrients. The *Urtica* (nettle) sub-community MG1b is especially frequent in areas of intensive arable agriculture. Where moister soils occur, in

Figure 2.18 Typical members of the *Arrhenatheretum* MG1 community: (a) cow parsley (*Anthriscus sylvestris*) and (b) dandelion (*Taraxacum* spp.).

ditches and at the bottom of slopes, the *Filipendula* (meadow sweet, Figure 2.19) subcommunity MG1c can occur.

The *Lolium-Cynosurus* MG6 community (Figure 2.17) is the major permanent pasture type on free-draining, circumneutral soils in lowland Britain. It forms the bulk of pasture for beef and dairy cattle. Subcommunity MG6a is widespread as a recreational sward on village greens, lawns and verges that are regularly mown. Tussocky patches of grass with lank rosette species (e.g. plantains like *Plantago major*) develop around dung pats, because animals do not like to graze near the dung. The so-called *avoidance mosaics* that develop constitute subcommunity MG6b.

○ From the examples given above, which group of grass-like plants would you expect to invade poorly drained areas of the field?

● Rushes (Juncaceae) are common plants of wet soils.

Poorly-drained hollows are often colonized by rushes (*Juncus* spp.) to form subcommunity MG10a, while near hedges and copses brambles (*Rubus* spp.) may invade to form subcommunity W24b.

The widespread use of cultivars of rye-grass (*Lolium perenne*) and the development of a range of styles of intensive grassland treatment have produced a wide variety of specialized grass-dominated and species-poor swards throughout lowland Britain, designated community MG7. The ribwort plantain (*Plantago lanceolata*) tends to invade resown areas that are subject to regular mowing or grazing and moderate trampling, to form subcommunity MG7e. Where trampling is very heavy and nutrient levels are raised, for example by frequent dog urination around gateways, the community tends to include grasses such as smooth meadow-grass (*Poa pratensis*) and very trample-resistant species like greater plantain (*Plantago major*) and daisy (*Bellis perennis*) to form subcommunity MG7f.

Figure 2.19 Meadow sweet is a characteristic member of MG1c subcommunity that flourishes in nutrient-rich, moist spots.

Question 2.3

Add an arrow below each entry in column 2 of Table 2.2 to indicate the direction of change in vegetation that results from the specified change in environmental conditions. The first row of the table is completed for you.

Table 2.2 Transitions in NVC community under the influence of environmental change.

Vegetation type	Nature of environmental change	Vegetation type
U5 *Nardus-Galium* grassland	increase in waterlogging and soil organic matter →	U6 *Juncus-Festuca* grassland
MG7e *Lolium-Plantago* grassland	decrease in trampling and nitrogen input	MG7f *Poa-Lolium* grassland
MG7 *Lolium perenne* leys and related grassland	decrease in intensity of cropping and fertilization	MG1 *Arrhenatherum elatius* grassland

Question 2.4

Match the terms describing spatial scale to the key factors affecting the distribution of grassland type at that scale.

Spatial scale terms: global, regional, local.

Key factors affecting grassland distribution: (a) soil type, management regime, microhabitat, slope; (b) climate; (c) soil pH.

2.4 Summary of Section 2

1   Most grasslands in Britain are semi-natural or intensive agricultural, and can be classified as either pasture or meadow, depending on their historical use. Most of them have evolved, since deforestation, and under the influence of shifting patterns of land management practices, into a wide variety of grassland types, some of which are an important reservoir of biodiversity.

2   The National Vegetation Classification (NVC) system can be used to describe the differences in species composition between grasslands, relating them to a range of abiotic, biotic and other influencing factors, such as soil type, hydrology and management practices.

3   The NVC system divides British grasslands on the basis of floristic composition, rather than function, into three main types: mesotrophic, calcareous and calcifugous/montane, a classification that relates to the pH of the soil. However, each of the three types is subdivided into many communities and subcommunities that are associated with a more specific suite of environmental conditions.

4   The distribution of grassland communities is spatially complex and highly dynamic. Because each plant species is characterized by a unique set of environmental tolerances (niche), a shift in one or more environmental variable can bring about a change in floristic composition.

5   An understanding of the functional relationships between NVC communities should enable us to develop sound policies and practices for conservation and restoration of important grassland habitats.

6   The variables that are primarily responsible for determining the types of grassland community depend on the scale of interest. Globally, climate is most important, but regionally, within Britain, soil pH is a key factor, while at a more local level many other variables come into play, especially nutrient supply, moisture, management and microclimate.

Grasslands: ecosystems in crisis?

3

Grasslands are grazing ecosystems in which herbivores and detritivores play key roles in determining not only the productivity and nutrient flux of the system, but also the species composition of the vegetation. Many grassland ecosystems have evolved as a result of long-established management regimes. Such semi-natural grasslands frequently have high wildlife value but, because of the complexity of the food webs and interactions with the environment, they tend to have low stability.

Human population growth makes increasing demands on the resources supplied by grasslands. Areas of grassland are lost through urbanization and conversion to arable land. Increases in (often illegal) hunting, and fragmentation of grasslands have a particularly detrimental effect on the larger, more wide ranging herbivores and predators, and this in turn has knock-on effects down the food chain.

Many of the problems that affect grasslands worldwide are exemplified by the situation in Britain. During the second half of the 20th century, 97% of British species-rich grasslands have been lost, mainly to development or agricultural intensification. Use of herbicides and pesticides has impoverished the native grassland flora and fauna. Application of fertilizers has also led to reduced diversity. Increasing mechanization and changes in the timing of cropping and tillage have removed sources of food and cover, with serious consequences for many wildlife species. Once destroyed, it is impossible to recreate the ecological complexity and diversity of an ancient grassland on a short time-scale. Restoration of these ecosystems takes decades, if it can be achieved at all.

3.1 Typical structure and dynamics of grassland ecosystems

As in any other ecosystem, a typical grassland ecosystem food web comprises:

- primary producers: the photosynthesizing plants such as grasses and forbs;
- primary consumers: herbivores such as bison, antelope and very often domestic grazing stock;
- secondary and tertiary consumers: predators such as lions, foxes;
- detritivores: such as dung beetles, millipedes, earthworms, and decomposers such as fungi and bacteria.

A difference between grassland ecosystems and other forms of terrestrial ecosystem is in the relative importance of the different components, that is, in the energy budgets. You learnt in Block 3, Part 2, that grasslands are typical grazing ecosystems.

○ Recall how the energy budgets of grazing ecosystems differ from those of storage and detritus ecosystems.

● Grazing ecosystems are characterized by the fact that much of the net primary production is eaten by herbivores, and the rest typically enters the detritus food chain when the plants die; very little of the primary production goes into storage as biomass or dead organic matter. In storage ecosystems a large proportion of primary production ends up being stored, and in detritus ecosystems less is eaten and more ends up as detritus.

3.1.1   Primary production in grassland ecosystems

Grasslands differ substantially from woodlands due to the relative absence of woody plants and the high biodegradability, and hence food value, of herbaceous material to grazing herbivores. On average, the total biomass for temperate grasslands is about 1.6 kg m^{-2}, of which at least 50% may be below ground, depending on grazing pressures and burning frequency (Table 3.1). Net primary production (*NPP*) is very variable, ranging from 0.2 kg m^{-2} in natural grasslands on impoverished soil to over 1.5 kg m^{-2} in artificially fertilized pastures.

Table 3.1   Net annual primary production and plant biomass for selected types of ecosystem.

Ecosystem type	*NPP*/kg m^{-2}		Biomass/kg m^{-2}	
	normal range	mean	normal range	mean
tropical rainforest	1–3.5	2.2	6–80	65
temperate deciduous forest	0.6–2.5	1.2	6–60	30
savanna	0.2–2.0	0.9	0.2–15	4
temperate grassland	0.2–1.5	0.6	0.2–5	1.6
tundra and alpine	0.01–0.4	0.14	0.1–3	0.6
arable land	0.1–3.5	0.65	0.4–12	1

○   According to Table 3.1, which type of ecosystem is generally more productive: grassland or forest?

●   Generally forests tend to have higher productivity than grasslands.

Yet most agricultural land, whether pastoral or arable, is based on exploiting grassland-type ecosystems.

○   Given their apparent lower productivity, why should agriculture be based on grassland rather than forest ecosystems?

●   Grassland primary production is the more digestible, so more of it can end up as secondary production. It is also quicker and easier to grow to maturity and harvest. The high productivity values for cultivated land show that the productivity of grass crops can be raised to that of forest.

It is the high palatability and high biodegradability of grassland vegetation that makes herbivory and the decomposer subsystem such dominant components of this sort of ecosystem.

3.1.2   The role of herbivores in grassland ecosystems

Selectivity

Irrespective of whether they are a natural part of the grassland ecosystem or domesticated animals on pasture, herbivores play important roles in the dynamics of grassland. Grazing animals tend to be selective in the material they eat, except in extreme situations where herbage is very limited, for example in prolonged drought conditions, after fire, or when stocking densities or population numbers are very high. Preferential grazing can influence the species, types of plants and their relative abundance within the community.

○ Recall the effect that selective grazing has on the MG6 NVC community when cattle avoid eating vegetation around a cow pat.

● The community changes to MG6b, in which rank, tussocky grass and lank rosette-type plants grow up around the cowpat, forming a characteristic 'mosaic' avoidance pattern.

Competitive species, which might otherwise achieve dominance in the sward, are constantly having their biomass removed, limiting their spread.

Nutrient flux

The presence of herbivores ensures the recycling of nutrients within a grassland ecosystem. Nutrients are returned to the soil in the form of dung and urine (Figure 3.1) and, in the case of unmanaged herbivore populations, dead animals. In pastures, nutrients are removed through the harvesting of milk, meat and wool. Patterns of dunging and the resulting nutrient enrichment depend on the wider abiotic environment, the type of herbivores present and any systems of management imposed on the grazers, for example moving, or 'folding', them to a specific area at night.

Figure 3.1 Nutrients are returned to the soil in the form of dung and urine.

Grazing enables faster cycling of nutrients than occurs in ungrazed grasslands. Plants are able to assimilate nutrients rapidly from dung and urine, producing fresh green material that is high in nutrients (especially nitrogen), in response. In ungrazed grasslands a high proportion of the nutrients within the system is locked up in the vegetation and soil organic matter. Some nutrients contained within urine and dung are lost through leaching or to the atmosphere, but much of the nitrogen, sulfur and potassium in urine, and the phosphorus, calcium and magnesium in dung, can be recycled and made available to plants. A study of one system showed that 80% of the nitrogen taken up by plants is recycled in the urine of grazers.

Trampling

Trampling is another important effect of large herbivores on grazing land. The impact of trampling depends on a number of factors, including herbivore density, type and size of animal, soil type, topography, length and season of grazing period, hydrology and rainfall. Trampling can be beneficial to the diversity of a grassland. The litter layer is broken down, coarse vegetation is broken up and gaps are created in which new seedlings can germinate. Areas of bare soil are essential in the life cycles of many invertebrates and may therefore contribute to the diversity of these ecosystems.

○ Recall from Block 3, Part 2, a butterfly species that is dependent on trampling of vegetation for the presence of its larval food plant.

● The high brown fritillary (*Argynnis adippe*) depends on trampling to provide enough light and space for the growth of its larval food plant, the common dog violet (*Viola riviniana*).

Where trampling is too concentrated or heavy, soil structure can be destroyed. The soil loses its ability to drain freely where hoof damage, combined with high soil moisture levels, destroys its natural porosity. This compaction or **poaching** of the ground (Figure 3.2) can result in loss of vegetation, causing soil erosion or invasion by short-lived species, usually regarded as weeds.

Figure 3.2   A heavily poached field.

Changes in grazing pressure may also exert an indirect influence on nutrient cycling pathways and ultimately on species composition of a sward. Figure 3.3 shows the hypothetical sequence of changes in nitrogen transformations resulting from reduction in grazing on a pasture ecosystem.

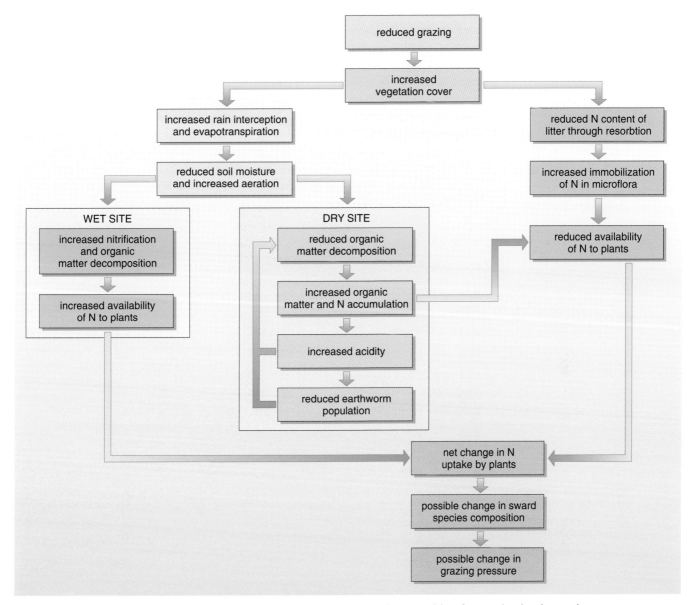

Figure 3.3 Hypothetical sequence of changes in nitrogen transformations resulting from reduction in grazing on a pasture ecosystem.

○ According to Figure 3.3, what is the major factor that determines the net effect of reduced grazing on nitrogen availability to plants?

● The moisture status of the soils.

At a wet site the availability of nitrogen is increased when grazing is reduced, but at a dry site the opposite is true. Either way, a change in species composition is likely.

Pollination

Large mammalian herbivores are such a visually dominant component of grassland ecosystems that it is easy to overlook the role of invertebrates. Their importance as detritivores has already been mentioned, and they are dealt with in more detail below (Section 3.1.3). Herbivorous insects, for example aphids, are generally thought of as pests of crops, but thousands of species of plants could not reproduce without their help. Around 90% of all flowering plants — including the great majority of the world's food crops — have evolved to depend on animals, especially insects, to transport pollen from one plant to another. Many other insects, birds and mammals are also dependent on either the seeds and fruits produced by pollination, or the pollinating insects themselves, as a food source. The importance of different pollinating agents is shown in Table 3.2.

Table 3.2   Pollinators of the world's flowering plants.

Pollinator	Number of plant species pollinated (approx.)	% of all plant species pollinated
wind	20 000	8.30
water	150	0.63
bees	40 000	16.6
all Hymenoptera (bees, wasps, ants, sawflies, etc.)	43 295	18.0
butterflies or moths	19 310	8.0
flies	14 126	5.9
beetles	211 935	88.3
thrips	500	0.21
birds	923	0.4
bats	165	0.07
all mammals	298	0.10
all vertebrates	1221	0.51

○   Which is the most important category of pollinator in terms of percentage of plants pollinated, and why do you think the total is greater than 100%?

●   Beetles are by far the most important category of pollinator (Figure 3.4). The total adds up to more than 100% because many plants are pollinated by more than one pollinator.

Invertebrates play a vital role in nutrient recycling processes in grasslands worldwide, many requiring the dung of large herbivores as part of their life cycle. This is discussed further in the following section.

Figure 3.4   Pollen beetle (*Oedemera nobilis*) eating pollen, UK.

3.1.3 The role of detritivores in grassland ecosystems

Detritivores play an important role in the recycling of nutrients and the overall health of grasslands. A broad spectrum of detritivores is represented in grasslands, including worms, millipedes, mites and beetles. In British grasslands we recognize the role of the common earthworm: turning over the layers of the soil and redistributing the nutrients from dead vegetation to humus. Dung beetles tend to be most readily associated with the savannas of Africa, yet they are in fact widespread throughout the grasslands of the world, with over 7000 species described.

○ Recall from the virtual field trip to the Teign Valley one species of rare British mammal that is heavily dependent on supplies of dung beetles.

● The greater horseshoe bat (*Rhinolophus ferrumequinum*).

African dung beetles act in a variety of ways: as 'dwellers', 'tunnellers' or 'rollers'. The 'dwellers' have the least evolved behaviour, simply taking up occupation in fresh dung, where the female constructs a nest for her offspring. The 'tunnellers' burrow into the dung and construct their nest chambers underneath it. They then push lumps of dung down the tunnels as a food source for their offspring. 'Rollers' are the most sophisticated type of dung beetle. They detach a piece of dung, in some cases up to 50 or more times their own mass, form it into a ball and then roll it away from the source of the dung and the attention of other beetles. These beetles lay their eggs in the dung ball, which they then bury. This act protects the developing offspring and provides them with a food supply, but also returns nutrients to the soil.

In West African savannas it has been estimated that dung beetles bury one metric tonne of dung per hectare per year. Primary production is significantly higher in savannas that are occupied by large herbivores and their associated dung beetles than in savannas where large mammals are absent. In Britain there are 56 *Red Data Book** species of beetle associated with different types of dung: 16 with cattle, 13 with sheep and 15 with horses, for example the Beaulieu dung beetle (*Aphodius niger*).

Question 3.1

Allocate the items from the list below to the appropriate rows in Table 3.3.

Lolium perenne, jaguar, high brown fritillary, detritivores, greater horseshoe bat, *Taraxacum* spp., aphids, decomposers, bison, sheep, spinifex grasses.

Table 3.3   Organisms and their role in a grassland ecosystem.

Role within grassland ecosystem	Examples of organism
primary producers	
primary consumers	
secondary and tertiary consumers	
	earthworm, dung beetle, millipedes
	bacteria, fungi

* For several groups of organisms, species in danger of extinction are listed in 'Red Data Books'.

Question 3.2

Select appropriate words from the following list to complete the passage below.

indiscriminate, consumed, digestibility, stored, dominant, toxicity, bees, predators, detritivores, beetles, selective, palatability, minor, flies, decomposed, 56, 12, 72.

Grazing ecosystems are characterized by the fact that little of the primary production is ———— . It is as a consequence of the high ———— and biodegradability of grassland vegetation that herbivores and ———— respectively are such ———— components of this sort of ecosystem. Herbivores can influence grassland ecosystems through their ———— grazing. ———— are by far the most important group of pollinators. In Britain there are ———— *Red Data Book* species of beetle associated with different types of dung.

3.2   Pressures on grassland ecosystems

Since the advent of agriculture, humans have been altering the landscape to grow crops and graze livestock, create settlements and pursue commerce and industry. As population sizes increased, so more land was converted to agricultural and urban development. Grasslands, especially, are of considerable economic importance, and as such are at great risk of loss, damage or fragmentation. Worldwide, humans have converted approximately 29% of the land area — almost 3.8×10^9 ha — to agriculture and urban or built-up areas, and grassland is the habitat that has suffered most from this.

You already know that, from a biological perspective, all grasslands are not considered equal. There are huge differences in their extent, distribution, biodiversity and stability.

○ Recall from Block 3 the two aspects to ecosystem stability.

● Ecosystem stability can be considered in terms of resistance to change and resilience: the ability to return to an equilibrium position after disturbance.

Problems arise because it is the rarest, most diverse types of grassland that are also the least resistant and resilient. In these grasslands, very complex and delicate food webs have evolved. Species are often completely dependent on each other and the regime of disturbance or management that perpetuates them. It can be very difficult to restore the ecosystem and ensure the survival of the dependent species where this balance has been upset, as the following example shows.

Case study: the large blue butterfly in Britain

In Britain, the large blue butterfly (*Maculinea arion*, Figure 3.5a) became extinct in 1979 and research, carried out during the efforts to reintroduce it, uncovered the exacting conditions needed for its survival. The butterfly relies on low-growing wild thyme (*Thymus polytrichus*, Figure 3.5b) as a nectar source, egg-laying site and food plant for the early larval stage. For its main period of growth the caterpillar is taken underground by *one* species of red ant (*Myrmica sabuleti*, Figure 3.5c) where it feeds on ant larvae until it is ready to form a chrysalis and, ultimately, to pupate. The ants are attracted to the butterfly larvae because they produce a sugary nectar, which the ants consume.

In order for the ants to survive in the grassland, the turf must be tightly grazed by herbivores. Even a small reduction in grazing can lead to a slightly taller sward and a cooling in microclimate to that favoured by a different species of ant, incompatible with the life cycle of the large blue butterfly. To achieve the exact balance in conditions has proved enormously difficult; early results of reintroduction programmes in southern England look promising, but the long-term future for the large blue remains uncertain.

3.3 Effects of increased pressure on grassland ecosystems

Figure 3.5 An eternal triangle? (a) The large blue butterfly (*Maculinea arion*). (b) Wild thyme (*Thymus polytrichus*). (c) Large blue butterfly larva being 'adopted' by a red ant (*Myrmica sabuleti*).

Clearly, urbanization leads to the loss of grassland habitat and the creatures therein, but fragmentation, hunting, and agricultural exploitation of grassland ecosystems also have adverse effects on their sustainability and biodiversity.

○ Why should fragmentation reduce the sustainability and biodiversity of some grassland ecosystems?

● Some animals, especially big predators and migratory herbivores that require large ranges, may be unable to survive if the extent of the habitat is reduced below certain limits. Increased gaps between similar areas of habitat may make dispersal for the purposes of breeding, migration or recolonization difficult or impossible.

3.3.1 Effects of hunting on biodiversity

The top predators and large herbivores of many natural grassland habitats have been hunted to near extinction. In the North American prairies the immense bison herds (*Bison bison*, Figure 3.6), estimated to have numbered around 60 million animals before the land was settled and developed, were largely destroyed by hunting. They are now being reintroduced to much of their former range.

Figure 3.6 North American bison were hunted to near extinction, but have been saved by conservation management.

One of the threats to the ecology of tropical grasslands is the extensive, and often highly organized, poaching of native animals. Skins, horns and tusks of various animals characteristic of these areas can command high prices, which threatens some species with extinction. Elephants, rhinos, snakes, big cats and birds have all been heavily exploited in this way; protection measures, export bans and wildlife reserves go some way to safeguard the future of these species, but their populations remain under threat.

3.3.2   The effects of conversion to agriculture

The biodiversity value of grasslands, whatever their type, tends to diminish as intensity of production or exploitation by humans increases. Once pastoral farmers move in to grazing lands, domestic livestock are introduced at the expense of native herbivores. On the more fertile land, native vegetation has been replaced by cereal crops, excluding many of the native fauna and flora. Furthermore, biological control experiments have sometimes had unpredicted consequences, and supposedly 'useful' introductions, such as the cane toad (*Bufo marina*, Figure 3.7) in Australia and the hedgehog on Orkney in Scotland have created havoc and ecological imbalance instead.

In Europe, the extent of agriculturally unimproved (more 'natural') grasslands has declined sharply in recent decades as a result of more intensive agricultural management. This change has been accompanied by a decline in species diversity. Figure 3.8 shows the trends in population numbers of some groups of British birds since 1970.

Figure 3.7   The cane toad (*Bufo marina*) was introduced to Australia to control pests, but without natural controls on its own populations it has now become a pest in its own right.

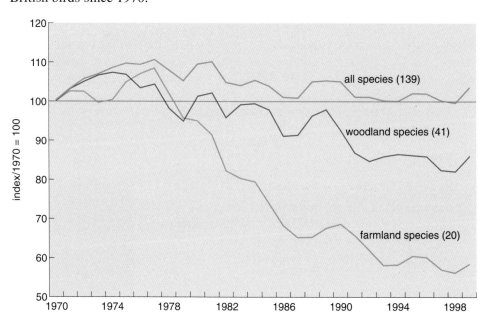

Figure 3.8   Relative numbers of wild birds in Britain, grouped according to habitat.

○   What does the graph show about the relative number of birds on farmland as opposed to birds in other habitats?

●   While populations of bird species in general have remained roughly stable, and woodland bird species have shown a roughly 10–15% decline since 1970, birds of farmlands have undergone a massive 40–50% crash in numbers.

Similar patterns of decline have been recorded for farmland plants and insects. The changes in farming practices that are connected with attempts to boost productivity include the introduction of new species and varieties of plant, the use of agrochemicals, such as fertilizers, pesticides and herbicides, changes to the pattern of tillage and cropping, and drainage of wetlands. In an effort to discover the underlying causes of these declines, organizations such as the Game Conservancy Trust and the Royal Society for the Protection of Birds (RSPB) have conducted experiments into the effects of the use of agrochemicals, and other changes in farming practice, on the abundance of selected species of native wildlife.

Farmers employ pesticides and herbicides to kill species of animals (mainly invertebrates) and plants (i.e. weeds) that threaten the production of crops. But, as you know, many of the invertebrate species on agricultural land are beneficial — as pollinators, consumers of pests (e.g. many beetles and spiders), and as food for wild birds such as grey partridge (*Perdix perdix*, Figure 3.9) and skylark (*Alauda arvensis*). Furthermore, many of the 'weeds' that farmers try to eliminate with herbicides are vital food plants for the adults or larvae of the beneficial insects. Indiscriminate use of these chemicals results in a dramatic loss of diversity that can unbalance the whole ecosystem. Modern insecticides commonly kill nearly 100% of the insects that would have provided food for partridge chicks.

In the past, hedgerows have also been important wildlife habitats in grassland areas (Figure 3.10a), providing a refuge for wild flowers and the insects that live on them. But the extensive removal of hedges to make room for larger farm machinery, and the use of herbicides right up to the base of the hedges themselves, has also reduced their value as wildlife habitat (Figure 3.10b).

Figure 3.9 Grey partridge (*Perdix perdix*).

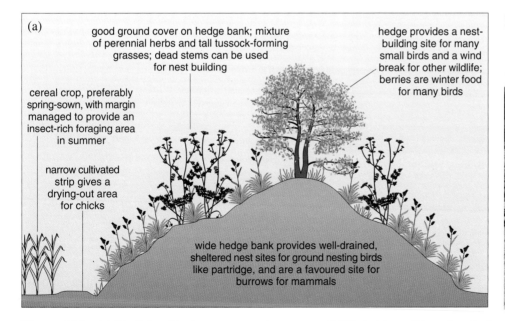

(a) good ground cover on hedge bank; mixture of perennial herbs and tall tussock-forming grasses; dead stems can be used for nest building

hedge provides a nest-building site for many small birds and a wind break for other wildlife; berries are winter food for many birds

cereal crop, preferably spring-sown, with margin managed to provide an insect-rich foraging area in summer

narrow cultivated strip gives a drying-out area for chicks

wide hedge bank provides well-drained, sheltered nest sites for ground nesting birds like partridge, and are a favoured site for burrows for mammals

(b)

Figure 3.10 (a) Traditional hedgerows were important wildlife habitats; (b) modern hedge management produces hedges of minimal wildlife value.

The loss of chick food, cover and winter forage are thought to have been major factors in the spectacular decline of the grey partridge and the skylark since 1970 (Figure 3.11). However, the skylark (*Alauda arvensis*, Figure 3.12), which is a ground nesting species, has suffered an additional threat from agricultural intensification. This species typically builds its nest in meadows. The shift from cutting for hay to cutting for silage has meant that mowing takes place earlier in the year than formerly. Consequently, many more larks' nests, and young larks, are destroyed.

Figure 3.11 Decline in numbers of the grey partridge and the skylark in Britain since 1970.

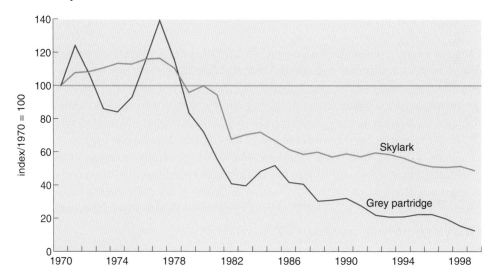

Most plants have their growth limited by the availability of major nutrients such as nitrogen and phosphorus. The application of fertilizers, which generally contain N, P and K, tends to boost primary production, often quite spectacularly. However, the species that typify our species-rich and increasingly rare unimproved grasslands, do not respond to high levels of nutrients. They have evolved to withstand naturally less fertile soils and have inherently lower growth rates. The species that do show rapid increases in growth rates when fertilized are those, such as ryegrass (*Lolium perenne*), that have been selected for use in commercial seed mixtures, for precisely that reason.

Figure 3.12 Skylark (*Alauda arvensis*), a ground nesting bird of farmland.

○ Using Grime's C–S–R scheme for classifying life history strategy, which strategies do the commercial seed mixture and traditional hay meadow species fall into?

● The commercial seed mixture species are competitors, whereas the slower-growing traditional hay meadow species tend to be more intermediate between stress tolerator and competitor.

Unfortunately, when high doses of fertilizer are added to a traditional unimproved hay meadow, many of the traditional species are likely to be eliminated by the few strongly competitive species.

Intensification of agriculture has also profoundly affected wet grasslands. Traditionally, grasslands that were fertilized by flood waters in winter were used either for hay or as summer pasture. They once provided habitat for large populations of waders and waterfowl. However, in recent years drainage of these seasonally wet lands has enabled farmers to use them throughout the year, causing the loss of wetland plant species and a range of threats to the wildlife. The use of fertilizers has led to the eutrophication of ditches and water channels (Topic 7). Populations of waders, such as the lapwing (*Vanellus vanellus*, Figure 3.13) have declined (Figure 3.14).

Figure 3.13 The lapwing (*Vanellus vanellus*) is a characteristic bird of wet grassland.

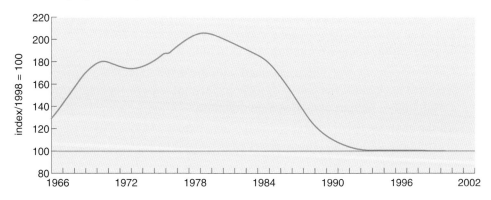

Figure 3.14 Relative abundance of breeding lapwing from 1966–1999.

The lapwing feeds on invertebrates that live in the soil, vegetation and pools of water that are found on wet grassland. Drainage of these grasslands has caused a reduction in invertebrate numbers and makes them more difficult to access. Lapwings nest on short grassland where, if stocking rates of cattle are too high, many nests are lost by trampling.

In most cases, threats to the ecology and biodiversity of grasslands around the world have arisen through exploitation of their productivity. In Britain, the main threats to unimproved grasslands have come from agricultural intensification, urban development, hydrological changes, succession to woodland and abandonment of traditional forms of management. In areas such as the Sahel, changes to grazing patterns have been exacerbated by many years of relative drought.

Global warming may pose many new challenges in the preservation and management of our valued grasslands in the future. Changing meteorological patterns could cause subtle or even radical changes to these habitats. Species distributions could alter and extinctions occur; new faunal species could invade ecosystems, and new communities could emerge. Only through careful and strategic monitoring will we be able to detect, and react appropriately to, any such changes.

Question 3.3

Using the information in Section 3, complete sentences (a) and (b) below:

(a) The four main types of pressure on grassland ecosystems are: ———, ———, ——— and ——— .

(b) The application of ——— to previously unimproved grasslands usually leads to the loss of plants with a ——— life history strategy, and an increase in ones with a ——— life history strategy (according to Grime's system), which usually leads to a ——— in species diversity.

3.4 Summary of Section 3

1   Grasslands are grazer ecosystems in which most primary production is either eaten by herbivores or goes straight to the decomposer subsystem. Storage of primary productivity is minimal. Grassland ecosystems are frequently complex and of low stability.

2   Large herbivores influence the floristic composition of grasslands through selective grazing, increased nutrient flux and trampling.

3   Invertebrates are an important part of the ecosystem, as herbivores, detritivores, predators and pollinators.

4   The main pressures on grassland ecosystems today come from urbanization, fragmentation, intensive hunting and agricultural intensification.

5   The management practices associated with the drive to increase agricultural productivity include the use of agrochemicals, increased stocking rates, destruction of hedges and effects on drainage.

6   All the changes listed in points 4 and 5 above tend to reduce the extent, distribution, variety and biodiversity of grassland habitats.

7   Climate change resulting from human activities may prove to be an even bigger threat to grassland ecosystems.

Grassland conservation management and restoration

4

4.1 Recognition of the need for remedial action

Over the years, the focus of conservation attention has changed. Globally, we are well aware of the threats to tropical rainforest and to high-profile species, such as the tiger and black rhino. The logo of the World Wide Fund for Nature (WWF) is a panda, one of the species that has tended to take the conservation headlines. However, many people are unaware of threats to other, less prominent habitats and 'unattractive' species, especially invertebrates. Until recently, local wildlife conservation often appeared to be contained within the bounds of nature reserves, cared for by specific conservation organizations, such as The Wildlife Trusts partnership in the UK. Increasingly now, more and more effort is going into extending good conservation practice to our farms, road corridors and other green spaces. Education and information are the key to encouraging people not only to value wildlife at both a local and a global level, but also to take responsibility for careful stewardship of the resources used. Today, much of the responsibility for education and conservation is devolved to small local organizations and interest groups.

The conservation value ascribed to communities may take into account a number of factors, some objective, others subjective. Species or habitat rarity can be quantified, and are often the main reason preservation measures are taken. However, if an area also has important historic or cultural links then a site may be seen as being of even greater conservation value, especially where traditional management has been maintained. Over more recent times, the aesthetic appeal of grasslands has become another important criterion for conservation. As our traditional 'flowery meads' have disappeared under the pressures of intensive agriculture or development, we have sought to preserve and recreate them (Figure 4.1).

Figure 4.1 Wild flower meadows are a target for conservation and creation.

At the Rio Earth Summit in 1992, a high level of biodiversity was hailed as one of the key indicators of success when assessing efforts to move towards sustainable use of the planet's natural resources.

Question 4.1

Select appropriate items from the list below to complete statements (a) and (b):

white rhino, great crested newt, tiger, skylark, ancient woodland in the New Forest, an overgrown churchyard in Hammersmith, an old hay meadow in Essex, heather moorland on Dartmoor

(a) Two animals that typify the early approach to conservation are ———— and ————, while two that typify a more modern approach are ———— and ————.

(b) Two habitats that typify the early approach to conservation are ———— and ————, while two that typify a more modern approach are ———— and ————.

4.2   Conservation of grasslands in Britain

Important British grasslands tend to be associated with a long history of consistent management, which has led to the establishment of communities recognized for their diversity or species richness.

○   Recall the meanings of the two terms, species diversity and species richness.

●   Species diversity encompasses number of species and their relative abundance, while species richness simply measures an overall list of species, with no account taken of their relative abundance.

In English lowlands, grasslands are associated with 25% of nationally rare or scarce plants, 20% of nationally rare or scarce birds and 65% of nationally rare or scarce butterflies.

Approximately 3% of all permanent grasslands in the English lowlands are of high botanical interest: probably less than 100 000 ha in total. Many protected areas of grassland are fragments of their original extent; they can be found throughout the British Isles, protected variously through statutory designation (e.g. as a Site of Special Scientific Interest, SSSI) or non-statutory status (e.g. as a local nature reserve). Some of the largest areas of protected grasslands of wildlife importance have survived under military ownership, for example Salisbury Plain (Figure 4.2) and Porton Down in Wiltshire. Others have survived by being part of common land, or have escaped the fashion for 'weed-free tidiness' on verges, in churchyards, at prehistoric earthworks, on cliffs or golf courses.

Figure 4.2   Many military training areas are important areas of grassland with great wildlife value.

Following the 1992 Earth summit, the Government published in 1994 its own strategy: *The UK Biodiversity Action Plan* (UKBAP). It proposed a partnership approach, combining existing conservation projects with new initiatives, aimed at conservation and enhancement for species and habitats, promoting public awareness and contributing to international conservation work.

Grassland habitats prioritized in the UKBAP are:

- coastal and floodplain grazing marsh (see Section 4.5)
- lowland hay meadow
- upland hay meadow
- lowland dry acid grassland
- calcareous grassland (see Section 4.4).

○ In what way do the criteria for this classification differ from those used for the NVC system?

● This classification is based on land use and location rather than the floristic composition used by the NVC system. However, each category can be translated into one or more NVC communities, and it is often the NVC methodology that is used to identify important sites.

In addition, rare plant and animal species associated with British grasslands, such as the great crested newt (*Triturus cristatus*, Figure 4.3), have conservation measures aimed at their protection and population increase.

Figure 4.3 The great crested newt (*Triturus cristatus*).

The great crested newt is associated with British grasslands mainly because of the small pools that are key components of British grassland ecosystems. Great crested newts use ponds for breeding, but also forage in the surrounding grassland for (mainly invertebrate) food at night, and hibernate on land during the winter.

Organizations such as the RSPB and the Game Conservancy Trust are involved in the conservation of habitats for particular species, while recognizing the interdependence of different biota within an ecosystem.

4.3 Management challenges

Having classified an area of grassland as deserving protection, one of the biggest challenges faced by those involved with conservation is that of determining effective and sustainable management practices. A comprehensive and strategic approach in the planning of management is essential, involving:

- gathering data on physical, biological and cultural factors;
- evaluation of the data;
- consultation with local interested parties;
- setting of clear and attainable objectives, both short- and long-term;
- prescribing precise management tasks and regimes;
- recording these and all other significant events; and
- monitoring and reviewing the plan.

One tried and tested model for grassland management decision making is shown in Figure 4.4.

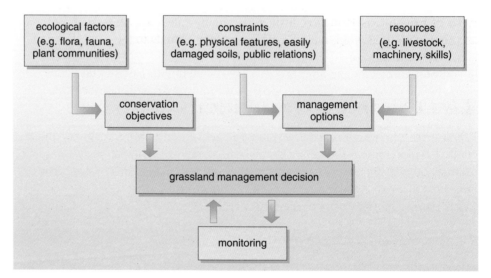

Figure 4.4 Simple model of grassland management decision-making.

○ Using the model in Figure 4.4, what factors affect the decisions made concerning the management of a particular grassland?

● In the first instance, the conservation objectives and management options influence the grassland management decisions, but these decisions can be modified later when the results of subsequent monitoring are taken into consideration.

The processes that once operated on traditionally managed, unimproved, semi-natural grasslands can be very complex and may be difficult to recreate. When it comes to restoring neglected sites, it is unlikely that a single management prescription will benefit the site as a whole. Therefore, without creating an over-complex system, a mosaic approach to management is often most appropriate; for example, while short turf may be necessary to maintain the plant interest of a site, taller vegetation is known to support a greater diversity of invertebrate species.

Management plans for these sites need to be drawn up carefully in full consultation with relevant conservation advisors. A change in management to more traditional techniques may have economic repercussions. Changes of water use within a catchment may have implications for the water regulating body, and alterations in drainage patterns may affect neighbouring landowners. Many aspects need to be considered in the management planning process. However, appropriate management is now being encouraged by government agencies responsible for conservation in Britain, for example the Countryside Stewardship Scheme, through advice and financial incentives to landowners.

Case studies provide us with a valuable insight into the successes and failures of putting theory into practice. In many ways, failures may prove more valuable than successes, as they can add to the understanding of complex ecological systems. However, despite the difficulties, there are some notable success stories to encourage conservation managers and planners. In the next two sections, two types of British grassland are considered, namely calcareous grassland and coastal and floodplain grassland, including their ecology, threats to their stability and conservation management.

4.4 Calcareous grasslands

When we think of calcareous grassland, images of close cropped, flower-strewn chalk downland invariably come to mind. In some turf a careful search may reveal as many as 40 different plant species within one square metre. The calcareous grassland that exists today represents just a fragment of what were, historically, extensive and integral parts of the farming system.

4.4.1 Historical background

Traditionally, calcareous grasslands were sheep-grazed by day, with the flock being returned to lower-lying areas of farmland at night where their dung would help to enrich the soil. At certain times of the year the sheep would be brought onto cereal stubble, brassica crops or water meadows, where they could feed prior to lambing. After lambing they would be returned to the downland, where the diverse herbage would supply all their nutritional needs.

It was not until the development of tracked vehicles, which enabled the extensive tracts of downland to be ploughed up, that these communities really came under any serious threat. Patterns of farming changed after the World War I, with further moves to convert land for arable use at the onset of World War II (Figure 4.5). Myxomatosis, the use of fertilizer and neglect caused other areas to lose their characteristic short turf and floral diversity. Today we value and conserve calcareous grassland.

Figure 4.5   A 'Dig for Victory' campaign poster from World War II.

4.4.2   'Important species': threats and causes of loss

Given the diversity of chalk grassland communities (the NVC recognizes 14 main communities), it is difficult to single out individual plant species as being of special importance. Many plants are so specialized that they are found only in the herb-rich turf over chalk or limestone. Chalk milkwort (*Polygala calcarea*, Figure 4.6a) and bastard toadflax (*Thesium humifusum*, Figure 4.6b) have limited powers of dispersal and are found only where there has been no major disturbance or change in management. Even where ploughing was last carried out up to 50 years ago, some species may not have been able to recolonize. Many of the characteristic plants of calcareous grassland are perennials and many individuals may be very long-lived. Grazing maintains the vigour of these plants, encouraging new shoots through regular defoliation.

Protected rare plant species prominently associated with calcareous grasslands include the military orchid (*Orchis militaris*, Figure 4.7a) and the Chiltern gentian (*Gentianella germanica*, Figure 4.7b).

Figure 4.6   Plants of calcareous grassland with poor powers of dispersal: (a) chalk milkwort (*Polygala calcarea*); (b) bastard toadflax (*Thesium humifusum*).

Figure 4.7   Rare plant species of calcareous grasslands: (a) military orchid (*Orchis militaris*); (b) Chiltern gentian (*Gentianella germanica*).

The butterfly species most commonly associated with calcareous grassland in England are the blues.

○   Recall the blue that relies on the red ant *Myrmica sabuleti*.

●   The large blue, *Maculinea arion*.

Of the many hundreds of insect species that are associated with calcareous grassland, many are rare and specialized, a fact that management plans must take into account. No bird species is confined to calcareous grassland; however, many species benefit from the food sources within the habitat. Green woodpeckers are known to attack anthills in search of larvae, kestrels and barn owls prey on small mammals, and skylarks and partridges feed on small insects and seeds. Wintertime sees flocks of birds such as finches, pipits and plovers gathering on the open grassland.

4.4.3 Management techniques

In most cases, grazing is the only satisfactory system appropriate to managing calcareous grassland over the long term. However, grazing can take many forms, which can have very different effects, so it's vital that appropriate choices are made during the planning stage if successful restoration is to be achieved.

Grazing, cutting and burning

Where grassland has been neglected and scrub has developed, initial remedial treatments may involve cutting (Figure 4.8), winter burning, or high-density grazing for a short period, or a combination of these approaches. Neither burning nor cutting should be used as a regular form of management as they both cause heavy losses to invertebrate populations. Grazing removes vegetation more gradually and enables the invertebrate communities to remain in balance.

Figure 4.8 Where scrub has developed, the rank vegetation may need cutting back.

○ Recall from Section 3.1.2 the beneficial effect that trampling has on vegetation.

● The physical action of the grazers' feet on the litter layer helps to break it up and creates bare patches where seedlings can germinate.

Choice of grazing animals

Occasionally cattle may be suitable grazers; however sheep are most commonly the best option as they nibble close to the ground without causing damage to the growing points or rootstock of the plants. Unlike cattle, sheep are light enough not to break up the turf or cause poaching of the soil. Cattle break down litter and tussocks more quickly than do sheep, and they also browse scrub in early summer. Ideally, however, cattle and sheep should graze in turn as the absence of a particular grazer helps to limit the build-up of parasitic worms and flukes in the soil.

Timing of grazing

Summer grazing is helpful in the reduction of coarse grasses and scrub, but it may cause a reduction in nectar supply if stocking densities are high. Winter grazing can be less harmful to invertebrates, many of which are dormant in the soil or in less accessible parts of plants; it also helps to create gaps for seedling colonization. Maintaining a low stocking density all year round avoids problems of overgrazing, erosion and poaching. A rotational grazing system, for example involving the use of flexi-netting (Figure 4.9) to create temporary enclosures, allows more flexibility to be built in around seasonal grazing requirements and the specific needs of invertebrate and plant species and habitats.

Figure 4.9 Orange flexi-netting is a useful way to create temporary enclosures, as shown for this habitat creation scheme at the London Wetland Centre.

4.4.4 Case study: Martin Down National Nature Reserve, Dorset

Martin Down National Nature Reserve is an area of 336 ha located on gently undulating land that overlies various drift deposits on top of chalk in Dorset. Grazing rights have existed for many centuries, and disputes over these rights led to neglect of the grassland from the 1960s. Myxomatosis caused a significant reduction in grazing pressure from the 1950s onwards. The northern part of the reserve was ploughed from World War II until 1957, when it was returned to grassland. It is estimated that the southern part has been under grassland for around 600 years. The land was purchased in 1978 by English Nature (EN) and Hampshire County Council, by which time the vegetation had become tall, less diverse and invaded by scrub. Many bird and butterfly species, such as the adonis blue (*Lysandra bellargus*), were becoming rare.

EN managed the restoration of the site. The principal aim was, and still is, to produce and maintain a mosaic of habitats including both grassland and scrub. This strategy should allow the important chalk grassland plants and habitats to flourish, as well as maintaining scrub habitats for bird species such as nightingales (*Luscinia megarhynchos*) and lesser whitethroats (*Sylvia curruca*, Figure 4.10), and invertebrates, including several declining butterfly species.

EN carried out selective scrub clearance by cutting and burning. They introduced their own sheep and encouraged commoners to exercise their grazing rights too. Electrified flexi-netting was used to enable a paddock system to be established over the whole site. The management regime involved a number of approaches. Scrub was cleared on a rotational basis from the more important grassland areas. The short, herb-rich grassland was subjected to a precise spring and autumn grazing regime, which allowed a fairly tall, open sward to develop during the summer.

Figure 4.10 Lesser whitethroat (*Sylvia curruca*) feeding chicks.

The results were very positive, with a tremendous increase in flowers and butterflies. Plant survey results showed an increase from an average of 18 species per metre square in 1979 to 30 species per metre square by the early 1990s. Strict calcicoles, including horseshoe vetch (*Hippocrepis comosa*, Figure 4.11) and orchids greatly increased in number.

The majority of longer grassland was winter-grazed annually, maintaining a moderately rich grassland that supports large populations of meadow browns and marbled white butterflies, as well as high densities of skylarks. In other areas the grassland was managed by rotational grazing one year in three, which maintains grassland cover throughout the year for invertebrates.

The scrub grassland mosaics, important for the Duke of Burgundy butterfly (*Hamearis lucina*, Figure 4.12) and the caterpillar food plant, cowslip (*Primula veris*), are more complicated to manage. This regime involves regularly cutting back about one-third of the scrub to prevent the canopy closing. In addition, grazing within the areas of scrub is encouraged to keep the sward open, but not at levels that would lose the important long grass–scrub transition zone.

Figure 4.11 Horseshoe vetch (*Hippocrepis comosa*).

In the years since 1978 the conservation management regime at Martin Down has demonstrated the possibility of restoring a neglected chalk grassland to something approaching its state prior to the cessation of management. Carefully derived objectives based on knowledge of the valuable ecological aspects of the site have enabled practicable prescriptions to be implemented, while balancing the interests of conservation, history and the local community.

Figure 4.12 Duke of Burgundy butterfly (*Hamearis lucina*).

4.5   Coastal and floodplain grasslands

While the full extent of coastal and floodplain grassland in the UK is not known, it is estimated that there may be a total of 300 000 ha nationally. However, the area still holding species-rich vegetation is less than 5000 ha. In these wet grassland habitats, the flooded fields provide overwintering grounds for huge numbers of waterfowl and wading birds, while in the summer months breeding wader species such as redshank (*Tringa totanus*, Figure 4.13a), lapwing and snipe (*Gallinago gallinago*, Figure 4.13b) nest in the lush grassland. The damp grassland and ditches of these seasonally flooded fields develop rich communities of flora and fauna.

Figure 4.13   Common waders of wet grassland: (a) redshank (*Tringa totanus*); (b) snipe (*Gallinago gallinago*).

4.5.1   Historical background

Floodplains form as rivers approach the sea, depositing nutrient-rich silt over the flat lands either side of the river channel. The fertility of floodplains make them a valuable grazing resource once the winter floods abate. Between 1960 and 1990 much grazing marsh was lost, mainly as a result of agricultural intensification.

○   Recall from Section 3.3.2 the major forms of agricultural intensification that have threatened wet grassland habitats.

●   Drainage, use of fertilizers and increases in grazing intensity.

4.5.2   Important species

Important species of wet grassland habitats, some of which may be under threat, include:

*   vascular plants such as fen orchid (*Liparis loeselii*, Figure 4.14), and penny royal (*Mentha pulegium*);
*   invertebrates, for example mole cricket (*Gryllotalpa gryllotalpa*);
*   birds, especially waders such as redshank, lapwing and snipe, and waterfowl, for example teal; and
*   mammals such as otter and water vole.

Many of the most important birds of wet grassland habitats are waders. These birds manage to coexist by having slightly different diets and nesting in slightly different types of vegetation.

○ Recall the term used to describe this method of coexistence.

● Differential sharing of the available resources within an ecosystem is known as *resource partitioning*.

Some of the differential requirements of three of the commonest waders to be found on wet grassland (lapwing, redshank and snipe) are shown in Table 4.1.

Table 4.1 Microhabitat preferences of waders breeding on wet grassland.

Habitat	Lapwing	Redshank	Snipe
high drainage channel water	+	+	+
soft soil	0	+	+
early season flooding	+	+	0
tall vegetation	–	0	+
species-rich vegetation	–	–	+
tussocks	+	+	+
rushes	0	0	+
grazing/mowing	+	–	–
trees/hedges	–	0	–

+, positive association; –, negative association; 0, no known association.

Figure 4.14 The fen orchid (*Liparis loeselii*) is a rare species of wet grassland habitats.

○ From Table 4.1, which aspects of microhabitat do all three species have in common?

● High levels of water in drainage channels and tussocks in the vegetation.

These three bird species feed on invertebrates, but their diets are subtly different, as shown in Table 4.2.

Table 4.2 Diets of three wetland waders.

Bird species	Principal food items	Proportion of diet	Location of food item
lapwing	earthworms	large	damp soil
	tipulid larvae	large	grassy vegetation (and pools)
	aquatic insect larvae	small	pools
redshank	earthworms	intermediate	damp soil
	tipulid larvae	intermediate	grassy vegetation (and pools)
	aquatic insect larvae	large	pools
	ragworms	variable	estuarine mud
	shore crabs	variable	shores
snipe	earthworms	majority	damp soil

○ On the basis of the data in Table 4.2, which of the three species would benefit most, and which least, from areas of standing water throughout the breeding season?

● Redshank would benefit most and snipe least.

Clearly, there is no single conservation management approach that will benefit all three species of wader. The best solution is to provide a mosaic of unflooded and winter-flooded grassland as well as permanent and temporary pools.

4.5.3   Summary of management techniques

Approximately 10 500 ha of floodplain and grazing marsh are protected by legislation. Other sites are under the sympathetic management of conservation organizations or local authorities. Land owned outside these categories may be included in one of the government-run schemes, such as Countryside Stewardship or the Environmentally Sensitive Area scheme where financial incentives are offered for appropriate management.

Water-level control

In broad terms, the most important aspect of management on these sites is water-level control, which may be achieved by several means:

- ditches, drains and pipes
- dams, bunds (large retaining banks) and sluices
- artificial movement of water, e.g. pumps.

These methods may be used to retain water over a whole site or in specific areas. Alternatively they can be used to assist in the surface drainage of the site. Although coastal and floodplain grasslands require occasional inundation by floods to maintain their distinctive character, the more diverse swards rely on efficient surface drainage afforded by a well-maintained drainage system. Water-level control is monitored so that adjustments to the regime can be made where necessary. Control of water levels is vitally important because it has such an impact not only on the type of vegetation that grows, but also on the invertebrate populations that are such an important part of waders' diets.

Without maintenance, ditches gradually silt up and become choked with vegetation, passing through a number of different successional stages on the way. Each successional stage has its own characteristic flora and fauna. **Slubbing** involves mechanical excavation of silted-up and overgrown ditches. The act of excavating is very damaging to wildlife, and newly cleared ditches have little wildlife value. Therefore the most wildlife friendly approach to management is a rotation of slubbing that disturbs only a fraction of the ditch system at any one time, and maintains ditches at a range of successional stages.

Additional management techniques

Water control methods can be combined with other modifications to management practices, such as modified grazing patterns to promote a mosaic of favourable vegetation types, and the use of nest protectors (wire cages that cover wader nests at risk from predators and trampling by stock).

Targeted monitoring of ecological parameters of the site, using key species as indicators of success, can be used to assess the success of new management regimes.

4.5.4 Case study: Holkham National Nature Reserve in Norfolk

Background

The Holkham grazing marshes were formerly saltmarsh. Starting in the 16th century, land was reclaimed by the construction of a number of sea-wall embankments and used for sheep grazing until World War II, when some areas were converted to arable use. By 1985 nearly half the area was under arable cultivation, and the wildlife interest of the marshes was extremely limited. The National Nature Reserve at Holkham was founded in 1967 and is managed by English Nature under agreement with the Holkham Estate. The grassland consists of three separate areas, each with independent hydrology (Figure 4.15). The grazing marshes lie in a matrix of other habitats: foreshore, dunes, arable farmland, saltmarsh and woodland. The main management objective has been to enhance the conservation interest over the 401 ha of permanent grassland.

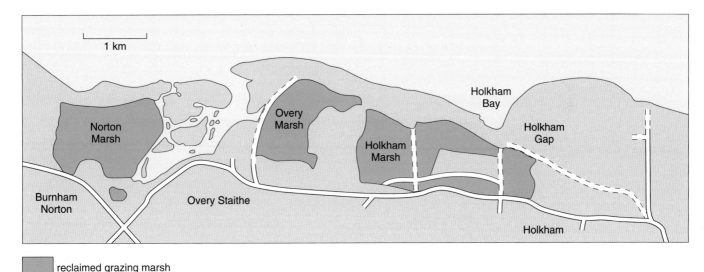

reclaimed grazing marsh

Figure 4.15   Map of the three wet grassland areas that make up the wet grassland part of Holkham National Nature Reserve.

Management approaches have included improved water-level control, grazing and mowing regimes.

Water management

Following a hydrological study in 1986 a number of measures were taken. Clay and timber dams and piped water controls were installed on Holkham Marsh. Existing dams were strengthened and a new timber sluice was constructed at Norton Marsh; piped water control points were constructed at Overy Marsh to ensure that neighbouring arable land was not flooded at high tides. A programme of ditch-slubbing with a seven-year rotation has been implemented.

By these means, partial flood conditions are created in winter on at least 10% of the areas and spring floodwater is retained for breeding waders and wildfowl, with at least 5% of the area being permanently flooded.

Grassland management

Grazing is limited to the summer, starting on the higher, drier fields (of lower conservation interest), and progressing downwards as the land gradually dries out. Low densities of cattle are used to achieve a sward length of 5–8 cm. Nest protectors are used to cover wader nests where necessary. On dry fields with little value for breeding birds, hay is cut from mid-summer and grazed thereafter. Silage cutting is being phased out. Use of fertilizers and herbicides is now limited.

Benefits

A mosaic of grassy fields, reed beds, shallow ponds, drainage channels and open water has been created. Water-table depths are higher and more varied than before. Drainage channels remain water-filled all year, which encourages aquatic plants and invertebrates. They also act as *wet fences*, which facilitates stock management. As a result of these changes, the vegetation and invertebrate fauna are much more diverse, and large numbers of birds have been attracted to the area. The numbers of wildfowl that overwinter on the reserve have increased almost tenfold in some cases; for example, average peak counts of teal have risen from 128 to 1460. Breeding waders and yellow wagtails have achieved a much higher breeding success rate (Table 4.3).

Table 4.3   Numbers of breeding birds at Holkham before and after conservation management was introduced in 1988.

Bird species	Numbers prior to introduction of management		Numbers after introduction of management	
	1986	1987	1990	1991
yellow wagtail	5	9	22	23
ducks				
mallard	12	19	48	61
tufted duck	0	2	8	11
shoveler	0	2	16	17
pochard	0	0	8	11
gadwall	0	1	9	12
waders				
lapwing	81	73	148	178
snipe	5	8	27	28
redshank	8	8	39	51

○   In terms of proportional increase, which wader and which duck have benefited most from the changes in management at Holkham?

●   Redshanks have increased proportionally more than the other waders (more than five times), pochard appear to be the most successful duck, as they were not recorded on the site before the changes to management.

As a result of the successful restoration the nature reserve is now designated a Site of Special Scientific Interest (SSSI).

Question 4.2

In Britain, what percentages of nationally rare or scarce plants, birds and butterflies are associated with lowland grassland?

Question 4.3

List the key components required of a model for a comprehensive and strategic approach to the effective and sustainable management of grasslands.

Question 4.4

(a) What national campaign initiated the ploughing up of grasslands during World War II?

(b) How many species might you expect to find, on average, within one square metre of herb-rich chalk grassland?

(c) How many calcareous grassland communities are recognized by the National Vegetation Classification?

(d) What are the advantages of sheep grazing over cattle grazing for most calcareous grasslands?

(e) What were the main causes of loss of grazing marsh from the 1960s onwards?

4.6   Summary of Section 4

1   In the past, the focus of conservation effort was on individual, high-profile species such as the panda. In recent years there has been a growing realization that many less charismatic species, in less dramatic habitats, merit conservation, and that there is a need to adopt a 'whole ecosystem' approach.

2   Criteria used to determine conservation status have included rarity, history and aesthetic value, but since the Rio Earth Summit of 1992 the emphasis has moved to biodiversity.

3   In Britain, government action has prompted the formulation of numerous action plans aimed at both species and habitats at a local and national level. A large number of organizations, governmental and non-governmental, large and small, has been involved in the formulation and implementation of these plans.

4   The grasslands recognized under the UKBAP are predominantly unimproved and semi-natural.

5   Several models for a comprehensive and strategic approach to the planning and implementation of effective and sustainable management for the conservation of grasslands have been developed.

6   A variety of management techniques have been used to restore both calcareous and wet grassland habitats.

Learning outcomes for Topic 9

After working through this topic you should be able to:

1 Explain what grasslands are. (*Question 1.1*)

2 Describe the extent of the global grassland resource and list the reasons why grasslands are important. (*Question 1.2*)

3 Explain the similarities and differences between natural, semi-natural and intensive/agricultural grassland types. (*Question 1.3*)

4 List and describe the principal environmental factors determining the occurrence of grassland types. (*Questions 1.3 and 1.4*)

5 Describe how the functional characteristics of the dominant vegetation types make them particularly well adapted to the grassland environment. (*Question 1.4*)

6 Name the grassland types from around the world that are quoted in the text, and give some examples of the key primary producers, herbivores and carnivores that characterize each type. (*Question 1.5*)

7 Describe briefly the historical influences that have affected the evolution of British grasslands. (*Question 2.1*)

8 When supplied with appropriate information, classify grasslands, using appropriate terms, according to the following criteria: climate, soil pH, soil moisture, human management and botanical composition. (*Question 2.2*)

9 Explain, using examples, the dynamic nature of community composition in space and time, and why community composition may change. (*Question 2.3*)

10 Explain, using examples, how the factors that determine the type of grassland (and other types of vegetation) vary as the spatial scale alters. (*Question 2.4*)

11 Describe, using appropriate examples, the structure of a grassland ecosystem and explain how such a system differs from other types of ecosystem. (*Questions 3.1 and 3.2*)

12 Describe and explain, using examples, the kinds of influence herbivores and detritivores can have on grassland ecosystems. (*Question 3.2*)

13 List the main types of pressures acting on grassland ecosystems today, and explain, using examples, how they can influence grassland composition, diversity and sustainability. (*Question 3.3*)

14 Describe how the focus of wildlife conservation has changed since the middle of the 20th century. (*Question 4.1*)

15 List the proportions of scarce plants, insects and birds associated with lowland grasslands in Britain, and the types of grassland prioritized by the UKBAP. (*Questions 4.2*)

16 List the key components of the model for a comprehensive and strategic approach to the effective and sustainable management of grasslands, and describe how it has been applied in specific case studies. (*Question 4.3*)

17 Name some of the characteristic flora and fauna of calcareous and wet grassland habitats, describe the historical background to their development, and explain, using appropriate examples, how the management techniques described can be used to restore them. (*Question 4.4*)

Answers to questions

Question 1.1

Grasslands are areas dominated by grassy vegetation and maintained by fire, grazing, drought or freezing temperatures. They encompass non-woody grasslands, savannas, and scrublands in which trees or shrubs are scattered at low density within the herbaceous vegetation.

Question 1.2

Roughly half the Earth's surface is occupied by grassland.

The principal reasons why grasslands are important are that they provide resources in the form of livestock (food, game, hides and fibre), and genetic resources. They also provide services in the form of employment, recreation, habitat, water purification and nutrient cycling.

Question 1.3

Column 1 of Table 1.2 should be completed as follows:

Grassland type	Characteristics
temperate	Often highly exploited and substantially altered through agriculture; they have become the main centres of world wheat production.
intensive/agricultural	Managed deliberately to produce forage crops and usually consist of a limited number of species.
tropical	Sometimes characterized by scattered trees or shrubs with a grass understorey; generally exist between the belts of tropical forest and desert vegetation.
natural	Distribution and species composition are determined by climate and soil type.
semi-natural	Communities representing deflected climaxes that are maintained by some human intervention, such as deforestation followed by regular burning or grazing.

Question 1.4

The most common functional type (from Raunkiaer's classification) to be found in grasslands is a hemicryptophyte. These types of plants are well suited to the grassland environment because their growing points are positioned close to the ground. This enables them to avoid damage caused by marked seasonality in terms of precipitation and temperature, as well as the damage caused by fire or grazing. Most grassland plants complete their reproductive cycle over a short period of time, and they must have stored reserves of energy to enable rapid shoot production, flowering and seed-setting.

Question 1.5

Columns 2 and 3 of Table 1.3 should be completed as follows:

Organism	Type of grassland	Community role
spinifex grass	tropical Australia	primary producer
white rhinoceros	tropical E. Africa	herbivore
marmot	steppe	herbivore
jaguar	pampas	carnivore
Schoenefeldia spp.	Sahel	primary producer
Stipa spp.	pampas or prairie or steppe	primary producer
N. American bison	prairie	herbivore

Question 2.1

At the time of the Domesday Book in 1086, it would appear that most of the large downlands were already in existence, whilst meadowland was common but not extensive. By the Middle Ages, large areas of grassland had been converted to arable use. The remaining grassland was more intensively utilized. More land was managed as meadow which had a value four times higher than arable land. During the 1300s, the Black Death drastically reduced the population; much arable land was converted to grassland. Later in the Middle Ages, a boom in the wool and cloth trade increased the need for sheep pasture. However, in the 18th century, the Agricultural Revolution wrought great changes to the land. Thereafter followed the Enclosure Acts, abolishing much common land. In 1954, an outbreak of myxomatosis had a severe effect on the short turf swards.

Question 2.2

The grassland is temperate, mesotrophic, wet, semi-natural, meadow, recognized as MG9.

Question 2.3

The directional arrows should have been added to Table 2.2 as follows:

Table 2.2   Transitions in NVC community under the influence of environmental change.

Vegetation type	Nature of environmental change	Vegetation type
U5 *Nardus-Galium* grassland	increase in waterlogging and soil organic matter →	U6 *Juncus-Festuca* grassland
MG7e *Lolium-Plantago* grassland	decrease in trampling and nitrogen input ←	MG7f *Poa-Lolium* grassland
MG7 *Lolium perenne* leys and related grassland	decrease in intensity of cropping and fertilization →	MG1 *Arrhenatherum elatius* grassland

Question 2.4

At a *global* level, factor (b) is most important in determining the distribution of grassland types.

At a *regional* level, factor (c) is most important in determining the distribution of grassland types.

At a *local* level the factors listed in (a) are most important in determining the distribution of grassland types.

Question 3.1

Table 3.3 should have been completed as follows:

Role within grassland ecosystem	Examples of organism
primary producers	*Lolium perenne*, *Taraxacum* spp., spinifex grasses
primary consumers	aphids, bison, sheep, high brown fritillary
secondary and tertiary consumers	jaguar, greater horseshoe bat
detritivores	earthworm, dung beetle, millipedes
decomposers	bacteria, fungi

Question 3.2

Grazing ecosystems are characterized by the fact that little of the primary production is <u>stored</u>. It is as a consequence of the high <u>palatability</u> and biodegradability of grassland vegetation that herbivores and <u>detritivores</u> respectively are such <u>dominant</u> components of this sort of ecosystem. Herbivores can influence grassland ecosystems through their <u>selective</u> grazing. <u>Beetles</u> are by far the most important group of pollinators. In Britain there are <u>56</u> *Red Data Book* species of beetle associated with different types of dung.

Question 3.3

You should have completed the sentences as follows:

(a) The four main types of pressure on grassland ecosystems are: <u>urbanization</u>, <u>fragmentation</u>, <u>excessive hunting</u> and <u>agricultural intensification</u>.

(b) The application of <u>fertilizers</u> to previously unimproved grasslands usually leads to the loss of plants with a <u>stress tolerator</u> life history strategy, and an increase in ones with a <u>competitor</u> life history strategy (according to Grime's system), which usually leads to a <u>reduction</u> in species diversity.

Question 4.1

You should have completed the statements with items (underlined) from the list, as follows:

(a) Two animals that typify the early approach to conservation are <u>tiger</u> and <u>white rhino</u>, while two that typify a more modern approach are <u>great crested newt</u> and <u>skylark</u>.

(b) Two habitats that typify the early approach to conservation are <u>ancient woodland in the New Forest</u> and <u>heather moorland on Dartmoor</u>, while two that typify a more modern approach are <u>an overgrown churchyard in Hammersmith</u> and <u>an old hay meadow in Essex</u>.

Question 4.2

In Britain, 25% of nationally rare or scarce plants, 20% of nationally rare or scarce birds and 65% of nationally rare or scarce butterflies are associated with lowland grassland.

Question 4.3

The key components of the model for a comprehensive and strategic approach to the effective and sustainable management of grasslands are: (i) data gathering (physical, biological and cultural), (ii) evaluation of data, (iii) consultation, (iv) setting of clear and attainable objectives, both short and long term, (v) prescribing precise management tasks and regimes, (vi) recording these and all other significant events, and (vii) monitoring and reviewing the plan.

Question 4.4

(a) The Dig for Victory campaign.

(b) Forty.

(c) Fourteen communities.

(d) Sheep nibble close to the ground but without damaging the rootstock; they are less likely to cause poaching of the soil.

(e) Agricultural intensification; drainage; increased use of chemicals including artificial fertilizer; peat extraction.

Acknowledgements for Topic 9 *Grasslands*

Grateful acknowledgement is made to the following sources for permission to reproduce material in this book:

Cover illustration: Mike Dodd/Open University; *Figures 1.1, 1.3, 1.9, 1.10, 1.12, 1.13, 1.16, 1.18, 2.2b, 2.14, 2.18a and b, 2.19, 3.5b, 3.13, 4.2, 4.3, 4.6a and b, 4.7a and b, 4.11, 4.13, 4.14*: Mike Dodd/Open University; *Figures 1.2, 1.5*: Stan Osolinsk/Oxford Scientific Films; *Figure 1.4*: 'The global extent of grasslands', (2000–2001) *People and Ecosystems: The Fraying*, World Resources Institute; *Figure 1.6*: Francois Gohier/Ardea; *Figure 1.7*: Cox, C. B. and Moore, P. D., *Biogeography: An Ecological and Evolutionary Approach*, 3rd edn, Blackwell Publishers Limited; *Figure 1.11*: Andrew Brown/Ecoscene; *Figure 1.14*: Martyn Colbeck/Oxford Scientific Films; *Figure 1.15*: Edward Parker/Oxford Scientific Films; *Figure 1.17*: Paul Franklin/Oxford Scientific Films; *Figures 2.1, 2.2a, 2.12, 2.13, 3.2, 3.10b, 4.1*: Hilary Denny/Open University; *Figure 2.3*: Copyright © The British Library; *Figures 2.5, 2.6, 2.7, 2.8, 2.9, 2.11, 2.16, 2.17*: Rodwell, J. S. (ed) 'Mesotrophic grasslands', *British Plant Communities — Grasslands and Montane Communities*, **3**, copyright © Joint Nature Conservation Committee published by Cambridge University Press; *Figure 2.15*: David Gowing/Open University; *Figure 3.1*: Kevin Church/Open University; *Figure 3.3*: Anderson, J. M. (1981) 'The structure and functioning of ecosystems', *Ecology for Environmental Sciences: Ecosystems and Man*, Cambridge University Press; *Figure 3.4*: Steve Hopkin/Ardea; *Figures 3.5a and c*: 'Large blue *Macuilea arion*', *The Millennium Atlas of Butterflies in Britain and Ireland* (2001) Martin Warren/Butterfly Conservation; *Figure 3.6*: Alan and Sandy Carey/Science Photo Library; *Figure 3.7*: Tim Halliday/Open University; *Figures 3.8, 3.11*: Reproduced from 'Breeding Birds', *The State of the UK's Birds 2000*, Copyright © 2000, Royal Society of the Protection of Birds; *Figures 3.9, 3.12*: Chris Knights/Ardea; *Figure 3.14*: 'Lapwing *Vanellus vanellus*', *Breeding Birds in the Wider Countryside 2000*, Report No. 252, British Trust for Onithology; *Figure 4.4*: Crifts, A. and Jefferson, R.G. (eds), 1999, *The Lowland Grass Management Handbook,* 2nd edn. Copyright © English Nature/The Wildlife Trust; *Figure 4.5*: Imperial War Museum; *Figure 4.8*: Alan Atkinson/BTCV; *Figure 4.9*: Copyright © Jo Treweek; *Figure 4.10*: Mark Hamblin/Oxford Scientific Films; *Figure 4.12*: Copyright © Alan Barnes Photography; *Figure 4.15*: Reproduced from 'Grassland areas in the Holkham National Nature Reserve', *Case Study 7, The Wet Grassland Guide*, Royal Society of the Protection of Birds.

TOPIC 10
TROPICAL FORESTS

Byron Wood

1	**Introduction**	**71**
1.1	What is a tropical forest?	71
1.2	Global extent of tropical forest	74
1.3	Deforestation in the tropics	75
1.4	Summary of Section 1	76
2	**Inside a tropical forest**	**77**
2.1	The biodiversity of tropical forests	77
2.2	Coexistence	84
2.3	Origins of high species richness and diversity in tropical forests	89
2.4	Tropical forest layers	91
2.5	Summary of Section 2	97
3	**Interactions**	**98**
3.1	Trophic levels	98
3.2	Food chains and webs	101
3.3	Nutrient cycling	104
3.4	Summary of Section 3	105
4	**The future of tropical forests**	**106**
4.1	Forest fragmentation	106
4.2	Reforestation	108
4.3	Conservation	109
4.4	Sustainable management	109
4.5	Gap analysis	110
4.6	Summary of Section 4	112
	Learning outcomes for Topic 10	**113**
	Comments on activities	**114**
	Answers to questions	**115**
	Acknowledgements	**117**

1 Introduction

Note: There are two activities in this topic that require you to access particular websites via the S216 eDesktop.

1.1 What is a tropical forest?

To someone who has never visited a tropical forest, the experience is never quite as expected. Your immediate feeling may even be disappointment. Expectations of a forest dripping with a rich myriad of flowering orchids, numerous monkeys and chattering parrots, are replaced with the reality of a wall of green woody vegetation of identical leaves, and not a single snake or monkey in sight, only mosquitoes. The forest reveals itself slowly, not all at once. If you are fortunate enough to spend time in a forest, you are rewarded daily with new plants and new insects, and encounters with the shy vertebrate fauna, which soon dispel those initial tinges of disappointment.

Imagine your first visit to Guyana's rainforest. As you approach the forest along the logging road, the Sun may be high in the sky and beating down on your back. The second you enter the forest, however, your world immediately changes. It's dark. The tall, towering trees form a tight canopy layer 30–40 m above your head, capturing most of the sunlight. It smells. There is an overriding earthy, musty smell of soil, dead leaves, and of rotting wood and fruit. The air is still, as there is little or no air movement under the forest canopy. It's noisy. In the daytime there is a general background soundtrack of insects humming and buzzing, with the occasional screech, hoot, grunt or howl from something larger and out of sight. As dusk falls, the noise level intensifies, as apparently every living thing in the forest starts shouting, competing with every other shouting creature to get heard over the din. If you spend the whole day in the forest, you will soon realize that, despite the dramatic differences between this world and the one you left behind, it's still very much a nine-to-five world. The forest has a precise daily rhythm and routine (and also a seasonal one). At dawn and dusk, or when rain is approaching, the howler monkeys call. At dusk, the roosting forest butterflies collect together, and the large owl butterflies start their aerial courtship flights. At precisely the same time, to the minute, each day, the leaf-cutter worker ants drop what they're carrying and return to their nest, like lines of workers downing tools and dispersing from the factory at the end of a shift. Party time. Just as some of us are day people and others night people, so it is with tropical forest animals. As the daytime species find a roost or nest for the night, the nocturnal species stir and prepare for a night of action.

Pristine tropical forests are referred to as virgin, undisturbed or **primary** forests, if they are thought to have been untouched by human influences such as logging. If forest has been logged, cleared or ravaged by fire, and then allowed to regenerate, the plant community and structure is very different to that of primary forest. It is then known as disturbed or **secondary** forest. Other classifications have been based on the relative closure of the forest canopy, and unsurprisingly are known as closed and open forest. Primary and closed forest is usually forest that is at the end of succession, and is known as **climax** forest. Like many plant communities, natural primary forest is never homogeneous in structure, even when there has been no logging or other artificial disturbances. Stretches of forest can be interrupted by gaps, which vary in size and frequency and are usually brought about by the fall of large trees. This can be as the result of death from old age or disease, wind, landslides, lightning strikes or encounters with large animals, such as elephants. In all mature forests, the populations of most tree species are present as young, mature and senescent (over-mature) individuals. This age profile has been referred to as the trees of the future, the present and the past, respectively. These groups of differently aged trees are not uniformly distributed and cause patchiness in forest structure.

○ By what process will secondary forest return to the primary or climax state?

● By the process of succession, which may take centuries (or even millennia) to complete.

The term 'rainforest' was first coined by the German botanist A. F. W. Schimper, in 1898, to describe forests that grew wherever annual rainfall was greater than 2000 mm. Schimper gave a brief diagnosis of rainforest as being 'Evergreen, hygrophilous [water-loving] in character, and at least 30 m high, but usually much taller, rich in thick-stemmed lianas [vines] and in woody as well as herbaceous epiphytes *'. This short definition, however, hides the multitude of forest types that make up the tropical forest biome (see Table 1.1 and Figure 1.1), and also gives a poor idea of what a tropical rainforest is, especially to those who are only familiar with the temperate forests of Europe and North America.

One of the outstanding and immediately apparent features of the tropical rainforest is that the overwhelming majority of the plants are woody and have the dimensions of trees. Although trees are the dominant plants of the rainforest community, most of the climbers and epiphytes are also woody. The undergrowth is also dominated by woody seedling and sapling trees, shrubs and young woody climbers. The only herbaceous plants are some of the epiphytes and these often

---

* An **epiphyte** is any plant that does not normally root in the soil but grows upon another living plant while remaining independent of it except for support. An epiphyte manufactures its own food by photosynthesis in the same way that other green plants do.

(a)

(b)

(c)

Figure 1.1   Examples of tropical forests: (a) lowland evergreen rainforest in Trinidad; (b) mangrove forest, Trinidad; (c) montane forest, El Yunque, Puerto Rico.

Table 1.1 The characteristics of various types of tropical moist forests.

Climate	Soil water	Soils	Elevation	Forest formation
seasonally dry	large annual shortage			*Monsoon forests*: various formations
	slight annual shortage			*Rainforests*: semi-evergreen rainforest
permanently wet (perhumid)	dry land	layered (i.e. a sequence of horizons); mainly oxisols* and ultisols†	lowlands	lowland evergreen rainforest
			1200–1500 m	lower montane‡ rainforest
			1500–3000 m	upper montane‡ rainforest
			3000 m to tree-line	subalpine forest
		podzolized* sands	mostly lowlands	heath forest
		limestone	mostly lowlands	forest over limestone
		very basic rocks	mostly lowlands	forest over very basic rocks
	water-table high (at least periodically)	coastal salt-water		beach vegetation
				mangrove forest
				brackish water forest
		inland freshwater	oligotrophic (low-nutrient) peats	peat swamp forest
			eutrophic (high-nutrient) soils	freshwater swamp forest
				periodically wet forest

* Described in Block 2, Part 2 *Earth*.   † Acid soils characteristic of tropical forests.   ‡ Found in mountainous regions.

represent only a small proportion of the undergrowth. Plant families that are low-growing herbs in temperate regions, are often woody and the size of trees in tropical forests. The botanist Richard Spruce (Figure 1.2) wrote the following account of the Amazon forest:

> Nearly every natural order of plants has here trees among its representatives. Here are grasses (bamboos) of 40, 60, or more feet in height, sometimes growing erect, sometimes tangled in thorny thickets, through which an elephant could not penetrate. Vervains form spreading trees with digitate leaves like the horse-chestnut. Milkworts, stout woody twiners ascending to the tops of the highest trees, and ornamenting them with festoons of fragrant flowers not their own. Instead of your periwinkles we have here handsome trees exuding a milk which is sometimes salutiferous [health-giving], at others a most deadly poison, and bearing fruits of corresponding qualities. Violets of the size of apple trees. Daisies (or what might seem daisies) borne on trees like alders.

Spruce, R. (1908) *Notes of a Botanist on the Amazon and Andes*, 2 vols, Wallace, A. R. (ed.), Macmillan, London.

Figure 1.2 Richard Spruce (1817–1893), one of the great 19th century European explorers of South America.

1.2 Global extent of tropical forest

In general terms, a tropical forest climate can be defined as one with monthly temperatures of at least 18 °C throughout the year, an annual rainfall of at least 1700 mm (and usually above 2000 mm) and either no dry season or one shorter than four consecutive months. The distribution of such climates over the Earth's surface is controlled primarily by the positions and intensities of the intertropical low-pressure belt and the subtropical anticyclones during the year, the east–west dynamics of the tropical atmosphere and the distribution of land and sea. In some areas, monsoonal wind circulations, the orientation of coasts and mountain ranges and the distribution of warm and cold ocean currents also play significant roles.

Tropical forests are found mostly in the humid equatorial belt between the Tropic of Cancer and the Tropic of Capricorn, at latitudes 23° N and 23° S respectively. Tropical forests today cover 6% of the Earth's land area, in three major regions: the American tropical forest in both South and Central America (the neotropics), the African tropical forests and the Indo-Malayan forests (the latter also referred to as Asia and Malesia and found from India to southern China and New Guinea). A fourth formation, much smaller than the others, is in northeastern Australia; although this region is sometimes lumped in with the Indo-Malayan forests, some authors argue that it should be treated as distinct. (See Figure 1.3.)

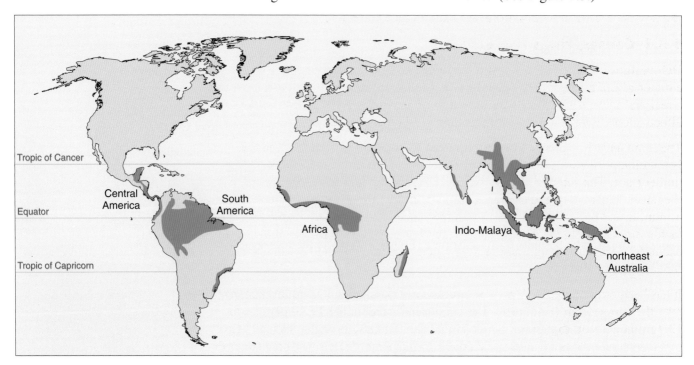

Figure 1.3 The main areas of tropical forest.

In terms of the total area of tropical forests that have a canopy cover greater than 10%, more than half (9 073 890 km^2) is found in the neotropical region (South and Central America); 91.2% of this area is in South America, most of it (67%) in Brazil.

Clues to the extent of the earliest tropical forests have come from a study of fossils. This type of evidence has shown that 70 million years ago there were forests present that were very similar to modern tropical forests. It is clear from the fossil record that flowering plants (angiosperms) began a phase of rapid evolution and many new species emerged at that time, including large broadleaved trees. It is amazing to think that so long ago, forests apparently very similar to tropical forests today, were found not only in the belt around the Equator, but also far outside the tropics, at much higher latitudes. Floral (i.e. plant) fossils from one of these forests have been described at latitude 60° N, in what is now known as Alaska. In both its floristic composition and its overall structure, this forest resembled a modern Indo-Malayan lowland rainforest. Out of 36 genera of trees found in the floral fossils of Alaska, 32 are now found in Malesia and seven are restricted to it. The similarity in structure and character of this Alaskan forest to a modern Malesian forest is even more remarkable. Leaf anatomy shows that the forest was overwhelmingly evergreen and 65% of the leaves had entire margins (i.e. a smooth edge, not serrated), and drip-tips (see Section 2.1.1) were common. These are all very characteristic features of tropical forest. It is also estimated that about a quarter of the species were lianas (vines and rope-like structures), another hallmark of tropical forests (Section 2.1.1).

It is generally accepted that as time progressed the climate became cooler, and by two million years ago it was no warmer in temperate regions than it is today. We can thus probably assume that when the subsequent glacial and interglacial periods began, the tropical forest had retreated to something more like its modern boundaries.

1.3 Deforestation in the tropics

1.3.1 Causes, rates and effects

Deforestation is the loss of forest cover. The rate of deforestation in the tropics is a major *global* concern. Although the effects of deforestation are felt at a local and a regional scale, mass tropical deforestation is also believed to have knock-on effects globally by altering weather patterns and ultimately climate.

The main underlying causes of deforestation of tropical forests are increased population pressures, a country's indebtedness and the structure of the world timber trade. The sources of deforestation are logging, fuel-wood collection, illegal settlement and land clearance for agriculture. These and other factors (such as mining and building) differ in their importance from region to region. The main factors implicated in the deterioration in the Congo's tropical forests in Africa, for example, were shifting agriculture, fuel-wood consumption and fires in open forests and savannas.

It has been estimated that 17 million hectares of tropical forest were lost per year over the period from 1981–1990. The estimated figure for the 1990–1995 period is 13.7 million hectares per year, which is still high, but does suggest that the rate of loss of natural forests in developing countries had slowed down compared with the decade before.

1.3.2 Measuring deforestation

In the past, mapping tropical vegetation relied on ground surveys. In the last 50 years, however, the detailed distribution of the tropical rainforest formations has become much better known. This is largely due to the increased availability of aerial photography and remote sensing techniques, such as radar imaging by Landsat geostationary satellites. Every 16 days, a Landsat satellite passes over the same point in the tropical landscape, recording digital data on the rainforests, savannas, rivers, lakes and mountains that lie 918 km below its orbit. These digital data are then converted into a false-colour image of different land-cover types, from which changes in land cover over time can be estimated. An advantage of using radar for remote sensing and vegetation mapping is that the wavelengths used (approximately 1 cm) are able to penetrate cloud cover, which is frequently present in the humid tropics. Another advantage of radar images is that they lend themselves well to electronic data processing, thus eliminating much of the subjective element involved in interpreting aerial photographs.

Activity 1.1 Satellite image interpretation

Landsat images covering the extent of the Amazonian rainforest can be found in the list of *Tropical Forests* links under 'Course resources' on the S216 course website. (http://www.grid.inpe.br/). The link will take you to a map of the Amazon Basin, which is part of the Global Resource Information Database, produced by the Instituto Nacional de Pesquisas Espaciais in Brazil. Describe what you think you can see in the following squares in the western (004, 63), central (228, 63 and 227, 63) and eastern areas (222, 63 and 221, 63) of the Amazon Basin. How easy is it to distinguish between different vegetation types (can you distinguish between different forest types?), and other features? What effect do you think this has on the accuracy of estimating rainforest cover?

1.4 Summary of Section 1

1   The majority of tropical forest plants are woody.

2   Pristine tropical forest is known as virgin, undisturbed or primary forest.

3   Forest that has been altered in some way and allowed to regenerate is known as disturbed or secondary forest.

4   In terms of succession, the primary forest plant community is known as climax forest.

5   Generally, a tropical forest climate is one with a mean temperature of 18 °C throughout the year, an annual rainfall of at least 1700 mm (and usually above 2000 mm) and either no dry season or a short one of fewer than four consecutive months with less than 100 mm of precipitation.

6   There are three major tropical forest regions; the neotropics (South and Central America), Africa and Indo-Malaya (the latter also known as Asia and Malesia), and a fourth smaller formation in Australia.

7   Fossil records show that some tropical forests from 70 million years ago were remarkably similar to some of today's tropical forests, in both floral (plant) composition and overall structure.

8   Deforestation is a major global concern.

9   Measuring deforestation is achieved through analysis of satellite images. The accuracy of interpreting these images determines the accuracy of measuring tropical forest cover.

Question 1.1

Summarize the similarities that have been found between floral fossils found in Alaska (thought to be 70 million years old) and forests today in Malesia.

Inside a tropical forest

2

2.1   The biodiversity of tropical forests

It has been estimated that the 6% of the world's land area that is covered by tropical forest contains 50% of its species. This global imbalance in occurrence of species is believed to be due to a combination of factors, but classically it has been explained by the fact that the tropics have optimal temperatures for growth and thus high productivity. As we shall see in this section, there are many other reasons for this higher biodiversity in the tropics, most of which increase the number of available niches for different species to occupy. The word 'biodiversity' is simply a contraction of the term 'biological diversity', and refers to the whole array of life-forms and their components, from genes, through species to habitats and ecosystems. Biodiversity, therefore, is all the genes, species and ecosystems inhabiting a region, and ultimately encompasses the number, variety and variability of life on Earth.

A region with a high biodiversity is often thought of as one having a large number of species. The number of species at a site is referred to as the species richness of that site. A site with a high species richness supports more species than a site with low species richness. The term 'species richness' is often used interchangeably with species diversity, although these two terms are distinctly different. Whereas species richness refers purely to the number of species, the term 'species diversity' also includes a measure of the number of *individuals* (abundance) of each species as well as the number of species. Species diversity measurements can therefore inform us of how evenly the numbers of individuals found at a site are spread between the total number of species found there. For example, suppose we record the number of tree frogs in two different one hectare tropical forest sites and find 20 species and 100 individuals, at both sites. Both these sites therefore have the same species richness. Now let's suppose that at one site, five individuals are recorded for each of the 20 species, and at the other site, two species account for 72 of the individuals, 10 species have two individuals each, and the other eight species have one individual each. The species diversity depends on how evenly the 100 individuals are spread between the 20 species. Clearly, in this fictitious example, the first site has a high species diversity and the second, a low species diversity. We will come across the species richness and species diversity of communities in more detail in the next section.

2.1.1 Flora

In a European or North American forest, the canopy trees usually belong to a few, sometimes only one, species. Even in the richest temperate forests such as the mixed broadleaved forests of China, the number of species is only 20–30. Tropical forests are very much more species-rich. On a single hectare of primary forest, there are seldom fewer than 40 (and often over 100) tree species with trunks over 10 cm in diameter. The explorer Sir Alfred Wallace (see Box 2.2, later) described this tree species richness in 1878:

Figure 2.1   The seed pods of the tamarind tree (*Tamarindus indicus*) of tropical Africa indicate its membership of the pea family (Fabaceae, subfamily Cesalpiniaceae). This tree is widely grown for the spice obtained from the pods.

If the traveller notices a particular species and wishes to find more like it, he may often turn his eyes in vain in every direction. Trees of varied forms, dimensions and colours are around him, but he rarely sees any one of them repeated. Time after time, he goes towards a tree which looks like the one he seeks, but a closer examination proves it to be distinct. He may at length, perhaps, meet with a second specimen half a mile off, or may fail altogether, till on another occasion he stumbles on one by accident.

Wallace, A. R. (1878) *Tropical Nature and Other Essays*, Macmillan, London.

Tropical forests are extremely rich in plant species. Of the 250 000 species of flowering plant in the world, approximately two-thirds are found in the tropics. Half of these are in Central and South America, 35 000 in tropical Africa, 40 000 in Asia and 25 000 in Malesia (see Figure 1.3).

There are similarities between the floras of the three main blocks of tropical forest, especially at the family level, but there are fewer genera in common and very few species. All three regions have many species belonging to the pea family (Figure 2.1), but half a dozen other families are well represented too.

Whereas America is characterized by numerous members of the Brazil nut family (see Section 3.3.2), with 11 genera and about 120 species, western Malesia is uniquely rich in members of the dipterocarp family (see Figure 2.2). Borneo, for example, has 287 dipterocarp species in nine genera, and in many places most of the big forest trees belong to this single family. Also, whereas there are many conifer species in Indo-Malaya, only one conifer species has been found in the lowland forests of Africa, and one other in tropical America.

Figure 2.2   A dipterocarp tree from Malesia. (Dipterocarp means literally 'two-winged fruit'.)

On a single 10 m × 10 m (100 m²) plot in wet lowland tropical forest in Costa Rica, 233 species of vascular plant and 32 species of bryophyte have been recorded. This is by far the richest plant community ever enumerated on Earth. By comparison, chalk grassland in England is extremely species-rich at a small scale, with 32–33 species in a 0.5 m² quadrat, but the total flora of one site of a few hectares is only 50–55 species.

To put this floral species richness into context, let us look at tree species richness in comparable plots of forest one hectare (100 m × 100 m) in size. The use of a standard plot size allows us to compare species richness data between sites. This within-community species richness has been called the alpha diversity (see Box 2.1) of each plot. In a one hectare plot at Yanamomo in the Peruvian Amazon, 283 tree species with a trunk at least 10 cm in diameter, have been recorded. Only 15% of these species had more than two individuals in this plot, and 63% had only one individual. In stark contrast, a one hectare plot in Nigeria contained only 23 species per hectare. A relatively rich temperate forest, on the other hand, may contain just ten tree species per hectare.

When sampling tree species richness in different forest sites, in the same country, there are other factors, apart from standard plot size, that should be taken into account to ensure comparability. Differences in some of the following factors may be entirely responsible for differences in the number of tree species found at different sites: altitude, local climate, local topography (are any sites on ridges, slopes or in standing water?), forest type (are they all lowland evergreen forest, or is one marsh forest?), forest disturbance level (perhaps several highly economic tree species have been logged and extracted from a site), and standardized sampling (if one survey has a minimum tree size of 10 cm in *diameter* at breast height, and the other uses 10 cm *girth* at breast height, the latter survey is including smaller and therefore more trees).

Box 2.1   Alpha, gamma and beta diversity

A count of species richness at a particular location is called the **alpha diversity** of that site. When a count is made of the species richness of a larger area, say a region, encompassing several communities, such as montane, cloud* and lowland forest, what we get is a measure of that area's **gamma diversity**. A third component of diversity describes the variation of habitats from place to place within the region. This is known as **beta diversity**. These three diversity measures are connected: gamma diversity = alpha diversity × beta diversity. In this way, we see that the total gamma diversity of a large area can be partitioned into two components, the local (alpha) diversity and the diversity of habitats, the beta diversity.

* Cloud forests occur above 1800 m and are almost permanently enveloped in cloud, with water droplets being constantly deposited on the foliage.

The State of Brunei in Borneo, South-East Asia, boasts about 2000 tree species. In contrast, the Netherlands, which is seven times larger in area than Brunei, has a mere 30 tree species. In fact, the whole of Europe north of the Alps and west of Russia has only 50 indigenous tree species and eastern North America has 171.

Not all tropical forests are as species-rich in terms of tree species as some of the above examples. There are naturally occurring tropical forest types where one or two dominant tree species may account for the majority of the individuals at a site. In seasonal evergreen forest in Trinidad in the neotropics, mora (*Mora excelsa*) trees can make up 85–95% of the trees forming the canopy (see Figure 2.3), and dominate all size classes apart from the 10–29 cm diameter classes (see Table 2.1).

Table 2.1   The percentage of *Mora excelsa* trees in different size classes in seasonal evergreen forest in Trinidad.

Diameter class/cm	*Mora excelsa* trees/%
10–19	13.6
20–29	13.5
30–40	35.4
41–60	54.6
≥61	91.3
all trees ≥ 10 and < 41	23.4
all trees ≥ 41	67.2

Figure 2.3   *Mora excelsa* forest (Trinidad).

Compare this dominance of the forest by one tree species (monodominance) with the Yanamomo site in Peru, where, as noted above, only 15% of the 283 trees had more than two individuals in a hectare and 63% were represented by just one individual.

Forests in which a single tree species is dominant are not a phenomenon exclusive to South America; they are also found in Africa. One such forest ranges from southwestern Nigeria, Cameroon and Gabon to the eastern frontier of the Congo and into northern Angola, and is dominated by the limbali tree (*Gilbertiodendron dewevrei*). In a sample hectare of this forest in the Congo, the limbali tree formed 94% of the trees that were equal to or greater than 20 cm in diameter at breast height. This forest covers hundreds of square kilometres — probably a larger area than any other type of monodominant tropical forest.

In wetter types of neotropical forest, the number of species of herbs, lianas, epiphytes and other 'non-trees' may exceed the number of tree species. At a site in Ecuador, 1033 species of vascular plant were recorded, of which only 154 (about 15%) were trees. The size and abundance of climbing plants, especially the woody lianas, is one of the most characteristic features of tropical forests. Lianas (Figure 2.4) use trees as support, and can often link trees so tightly together that even when a tree has been cut clean through at the base, it remains standing. It has been estimated that more than 90% of all species of climbing plant come from the tropics. A liana's strategy is to reach the well-illuminated canopy with great economy of stem material and then concentrate its resources on the production of leaves and flowers. Lianas are such formidable competitors with trees for space and light, that they often cause the crowns of trees to become misshapen or one-sided. It has also been found that the liana load has a highly significant negative effect on the growth in diameter of the host tree.

○   Describe what this negative effect, of liana load on the diameter growth of the host tree, actually means.

●   It means that the greater the number or biomass of lianas found on a tree, the slower the tree's trunk grows, and hence eventually, the smaller the diameter of that tree, compared with others of a similar age with a smaller liana load.

In large gaps, blankets of climbers — which may persist for several years — often make it difficult for tree seedlings and saplings to survive. This strong competition between lianas and trees has resulted, not surprisingly, in trees having evolved defence mechanisms against lianas. The small West African tropical forest tree *Barteria fistulosa* has a mutualistic relationship with ants of the genus *Pachysima*. One of the important beneficial relationships for the tree is that the ants bite off the tendrils and growing points of climbers that try to attach themselves to their host. The ants also attack other plants in the neighbourhood so that bare circular areas are formed around the trees that have been successfully colonized by these ants.

Strangler plants begin life as epiphytes and later send down roots to the soil, becoming independent, or almost independent, and often kill the trees by which they were originally supported. The strangler figs, in the *Ficus* genus, start off as germinating seeds on tall trees, often in a fork between

Figure 2.4   Lianas in a tropical forest.

the trunk and a large branch. They have been deposited there, usually by birds. The seedling grows into a large bush, which sends down aerial roots. These roots eventually reach the ground and embed themselves in the soil. The roots multiply and branch to form a mesh that engulfs the original host tree. The large crown produced by the strangler fig outcompetes its host for light, and eventually the host tree dies (Figure 2.5).

Epiphytes grow attached to tree trunks and branches. In closed forest, most epiphytes grow high above the ground, where there is strong illumination. Their only sources of nutrients are atmospheric precipitation and dust, the small quantities of solutes leaking from their host tree, and organic debris which they accumulate in various ways. Most

Figure 2.5 A young strangler fig tree (on the left) using a palm tree as its host.

epiphytes have to cope frequently with water stress, due to high rates of evaporation in between rain showers. They therefore have to absorb water quickly when it is available and to conserve it when it is not. Epiphytes provide the chief habitat for a rich and varied fauna, which plays an important part in the forest ecosystem. They provide the main nesting site for many species of arboreal (tree-dwelling) ants. The masses of humus collected by these ants have a large associated fauna of small invertebrates, and the large quantity of water collected in the overlapping leaves of a bromeliad is home to many organisms, as we will see later (in Section 2.2). Bromeliads are members of the Bromeliaceae family, mainly from the Americas, where they are most frequently found as epiphytes.

In terms of its flora, Africa has been called 'the odd man out'. There are fewer families, genera and species in African tropical forests than in either America or Indo-Malaya. For example, one mountain in northern Borneo, Mount Kinabalu (4100 m), has the same number of fern species as the whole of the African continent. Africa also has very few palm genera and species compared with the other main areas of tropical forest. In mainland Africa, there are only about 60 palm species, whereas in both tropical America and Malesia, there are over 1000.

Trees of tropical forests have three anatomical features that are much more common throughout these forests' global range than in other forest types: buttresses, cauliflory and drip-tips.

Buttresses are the wing-like expansions at the base of a large proportion of the bigger tree trunks (Figure 2.6). This trait of buttressing does not belong to any one family or group of families. The adaptive significance of buttresses may be mechanical, in terms of support, reducing the risk of being blown over by strong wind.

Figure 2.6 Buttressing on a large rainforest tree.

Figure 2.7 An example of cauliflory (*Swartzia* sp.).

Many tropical forest trees, especially smaller ones, bear flowers directly on the trunk or on its larger branches, a feature known as **cauliflory** (Figure 2.7). Wallace suggested that cauliflory was a device by which smaller forest trees, excluded from the abundant light and space of the canopy, could display their flowers where they could be easily seen and visited by shade-loving insects.

Another feature common to all tropical forests is that the leaves of many of the trees that make up the **understorey** (the layer between the forest floor and the canopy) have an exaggerated pointed end or **drip-tip**, so called because it was originally thought to help water drain rapidly from the leaf surface. The adaptive significance of drip-tips has fascinated scientists for over a century and various theories have been proposed and rejected. It has been suggested that leaves with a drip-tip dry more quickly after rain and those leaves without drip-tips have more algae and bryophytes covering them, which might interfere with photosynthesis. Others felt that having a wet leaf surface would lower the leaf temperature and depress the rate of transpiration (although this might also slow down the uptake of minerals). Other authors have rejected these ideas and proposed new ones in their place. It may simply be that drip-tips have *no* adaptive significance and are simply the result of the conditions under which the young leaves develop.

2.1.2 Fauna

As well as commenting eloquently on the tropical forest flora, Sir Alfred Wallace (Box 2.2) also wrote the following, which encapsulates the wonderment of many naturalists at the extraordinary *faunal* richness of tropical regions:

> Animal life is on the whole, far more abundant and varied within the tropics than in any other part of the globe, and a great number of peculiar forms are found there which never extend into temperate regions. Endless eccentricities of form and extreme richness of colour are its most prominent features, and these are manifested in the highest degree in those equatorial lands where the vegetation acquires its greatest beauty and fullest development.
>
> Wallace, A. R. (1878) *Tropical Nature and Other Essays*, Macmillan, London.

Examples of the high species richness of tropical forests come from many taxonomic groups. For instance, more than 1000 amphibian and 1100 reptile species have been recorded in South America. The means by which so many amphibian species can occur together is examined later (see Section 2.2). Approximately 500 reptile species are known from the neotropical lowland forest, 300 of which are endemic to this region. The richest river basin in the world in terms of freshwater fish is the Amazon Basin (Amazonia), with over 1300 species. As with the flora, this species richness does differ between tropical forest regions, however. In terms of the avifauna (birds), 3300 neotropical species have been recorded. Of these, 1300 are forest species, compared with 800 forest bird species in South-East Asia and 400 in Africa.

Box 2.2 Alfred Russel Wallace

After the publication of *The Origin of Species* in 1859, evolution by natural selection — biology's great unifying concept — became known as 'Darwin's theory'. First announced jointly the previous year, it is actually the Darwin–Wallace theory, but due to Darwin referring to it as 'my theory' and Wallace's huge modesty, evolutionary science is known today as Darwinism instead of Wallacism. This modesty was exemplified when, in 1889, Wallace titled his own work on evolution 'Darwinism'.

As a young school teacher in Leicester, Alfred Wallace (Figure 2.8) met Henry Walter Bates, who shared his passion for natural history. On weekend bug-collecting jaunts, the would-be adventurers discussed such favourite books as Charles Darwin's *The Voyage of the Beagle*, and dreamed of exploring the lush Amazon rainforests. Whereas Darwin's expeditions were funded by his father, Wallace's achievements are the more remarkable, for he financed his entire expedition to the Amazon (1848–1852) by selling thousands of natural history specimens, mainly insects, for a few pence apiece.

After four years of collecting and exploring, and having made numerous discoveries (despite malaria, fatigue and the most meagre supplies), Wallace boarded a ship to return to England. With him went his precious notebooks and sketches, an immense collection of preserved insects, birds and reptiles, and a menagerie of live parrots, monkeys and other creatures.

On the way home, however, his ship caught fire, and as he dragged himself into a lifeboat he was able to rescue only a few notebooks. The measure of Wallace's courage and resilience was shown shortly after his return to England, when immediately after receiving insurance money for the loss of some of his collections at sea, he embarked on a new expedition to the Malay Archipelago (1854–1862).

As well as being the co-creator of the theory of natural selection, Wallace discovered thousands of new tropical species. He was the first European to study apes in the wild and author of some of the best travel books ever written, including *Travels on the Amazon and Rio Negro* (1869) and *The Malay Archipelago* (1872). Among his remarkable discoveries is 'Wallace's Line', a natural faunal boundary in Malesia (now known to coincide with a junction of tectonic plates) separating Asian-derived animals from those that evolved in Australia.

Figure 2.8   Alfred Russel Wallace (1823–1913).

Activity 2.1   Quantifying diversity

The current herpetofauna (amphibians and reptiles) species list compiled by The Biological Diversity of the Guianas Program (BDF) is thought to represent 70% of the actual species list for Guyana. It can be found in the list of *Tropical Forests* links under 'Course resources' on the S216 course website. (http://www.mnh.si.edu/biodiversity/bdg/guyherps.html). Assuming every entry on the list is a different species, what is the species richness of the Hylidae family of frogs? Do you think this value represents the alpha or the gamma diversity of this family?

These examples of vertebrate species richness almost pale into insignificance, however, when insect species numbers are explored. In terms of the number of described species of all plants and all animals, 57% are insects, with the beetles representing 25% of all described species. It has been estimated that perhaps as many as 75–90% of all insect species are confined to tropical moist forests. As many as 67–99% of all insect species have yet to be described and given a name, and many await discovery. How many will have been lost before they have been described?

To give an example of how much greater this insect species richness is in tropical regions compared with temperate ones, for a well described group such as the butterflies, look at Table 2.2.

Table 2.2   Comparison of butterfly species richness in temperate and tropical countries and regions.

Country/Region	Area/10^3 km^2	Number of species	Species per 10^3 km^2
UK	244	58	0.24
Europe	10 500	358	0.034
Trinidad	5	630	126
Venezuela	910	2316	2.5
neotropics	20 000	7180	0.36

Having been given a flavour of both floral and faunal species richness in the tropics, we will now go on to explore some of the mechanisms and processes that might explain this overwhelming abundance of biodiversity in tropical forests.

2.2   Coexistence

We will now examine how so many species can exist together in the same habitat, apparently all at the same time and in the same space. Theories on the origin and maintenance of floral and faunal diversity in tropical forests will be explored in Section 2.3. The 'partitioning' of diversity to allow coexistence in the same forest is down to various forms of specialization, which can be temporal, spatial or resource-based (both for food and breeding sites), and is frequently a combination of these. Specialization takes the form of partitioning resources, such as space or time, which allows coexistence through reduced competition between individuals. For example, if some individuals carry out all of their activities in the forest canopy, and others carry out all of theirs in the understorey, then there is little or no overlap between individuals from these different parts of the forest, and little or no competition. We are exploring here why so many different species can coexist, without running out of resources such as food and space. We are therefore looking at the competition between individuals of different species, which is called interspecific competition.

○ Recall from Block 3, Part 2 *Life*, the name given to the idea that two species *cannot* coexist if they use the same set of resources.

● The competitive exclusion principle.

2.2.1 Temporal partitioning

One of the most obvious strategies developed to avoid competition and allow coexistence is temporal specialization, where some groups are active by day (**diurnal**) and others by night (**nocturnal**). Examples of temporal partitioning are found in many groups including the Lepidoptera, which are split roughly into the day-flying butterflies and the night-flying moths, although a few brightly coloured day-flying moth species do exist.

2.2.2 Spatial partitioning

The bird community at La Selva forest in Costa Rica is separated spatially, with five different levels of vertical zonation (see Table 2.3).

Table 2.3   Spatial partitioning of birds at La Selva, Costa Rica.

Above canopy	vultures and swifts
Canopy top	toucans and parrots
15–25 m	woodpeckers and jacamars
Understorey	most hummingbirds, antbirds and manakins
Forest floor	great currasow, ground doves and wrens

Spatial partitioning between layers of the tropical forest is also illustrated by the different niches of rainforest mammals (Figure 2.9) and is explored in more detail in Section 2.4.

Spatial specialization can also be found at a much finer scale than the coarse spatial scale of different layers in the tropical forest. The epiphytic bromeliads of the neotropics provide a very specialized niche for many species (Figure 2.10). The overlapping leaf bases of the epiphytes in the Bromeliaceae family form a reservoir, which in a large bromeliad can collect as much as five litres of rainwater, as well as organic debris. These 'tanks' of water are home to a multitude of organisms, including freshwater crabs and amphibians. In Jamaica, 68 species have been found to inhabit these aerial aquaria, including freshwater crabs, tree frogs (Figure 2.11), and insect larvae such as mosquitoes. In turn, this bromeliad 'zoo' attracts predators such as lizards, scorpions and snakes.

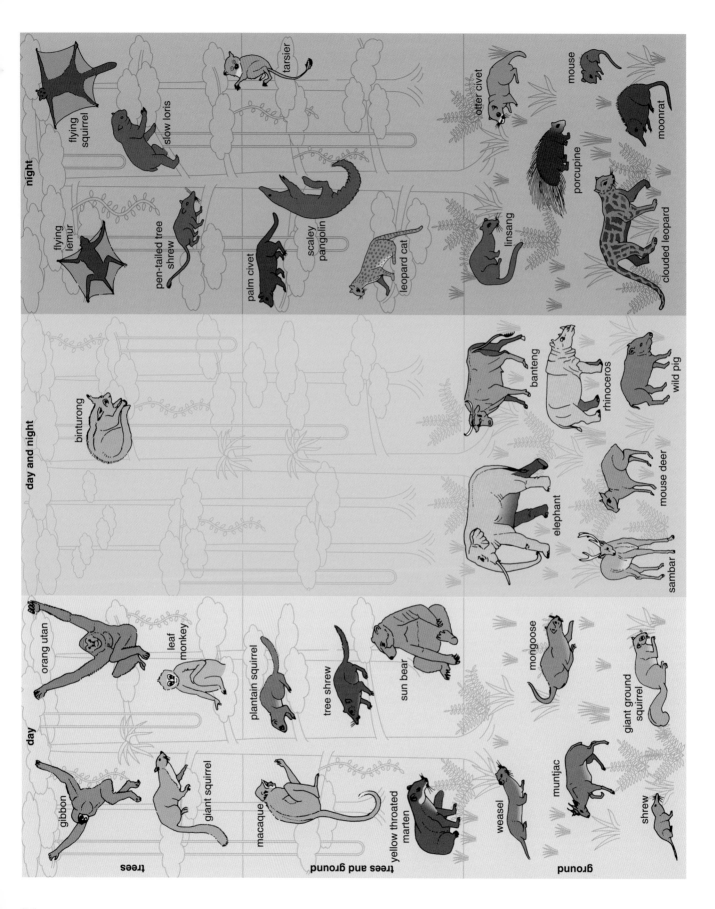

◀ **Figure 2.9** Spatial and temporal partitioning of non-flying mammals in the lowland rainforest of Sabah, Indonesia.

Figure 2.10 A bromeliad from the neotropical Bromeliaceae family, found in Trinidad, showing the water traps formed by the overlapping leaf bases.

Figure 2.11 The female of this Trinidadian tree frog (*Flectonotus fitzgeraldi*), at 25mm in length, squeezes into the bromeliad water traps to lay her eggs.

The pitcher plants (*Nepenthes*) of the Indo-Malayan tropics provide another special niche for animals, with 55 insect species being found as inhabitants, two-thirds of which live and breed there.

2.2.3 Food partitioning

The hummingbirds are generally small, brightly coloured birds belonging to the Trochilidae family, and with 343 species, they make up the Western Hemisphere's second largest family of birds (Figure 2.12). Spending the majority of their day in flight, with an average wing-beat rate of about 50 per second and a heart rate of 1260 beats per minute, hummingbirds require large quantities of both energy and protein. They have been recorded visiting up to 1000 flowers a day to feed on floral nectar, and supplement this with insects as a source of protein. Hummingbirds have evolved to feed on individual species of flowers. Species have evolved specific beak sizes and shapes to allow feeding on the nectar of flowers of similar shape and size. In a forest community of hummingbirds, the food resource of floral nectar is thus shared between the community, so reducing interspecific competition.

Figure 2.12 Different hummingbird species feeding on flowers of different shape and size: (a) the Bahama woodstar (*Calliphlox evelynae*) at *Hibiscus rosasinensis*; (b) the snowy-bellied hummingbird (*Amazilia edward*) at *Cuphea* sp.

(a)

(b)

In northeast Gabon in Central Africa, there are five nocturnal species of lorises (primitive primates) that exist in the same habitat. Of the three species that live in trees, the bushbaby *Galago* spp. (Figure 2.13a) is mainly insectivorous, another bushbaby (*Euoticus elegantulus*) feeds on plant gums and the potto (*Perodictus potto*, Figure 2.13b) eats mainly fruits. The two undergrowth species are a fruit-eating bushbaby (*Galago allenii*) and an angwantibo (*Arctocebus calabarensis*, Figure 2.13c), which eats insects.

(a)

(b)   (c)

Figure 2.13   (a) A bushbaby (*Galago moholi*), (b) a potto (*Perodictus potto*) and (c) an angwantibo (*Arctocebus calabarensis*).

○   What type of resource partitioning allows these five nocturnal loris species to avoid competition and coexist in northeastern Gabon?

●   A combination of food (resource) and space (spatial) partitioning.

2.2.4 Breeding sites

Another strategy that has evolved to allow species to live together is demonstrated by the rich frog and toad communities of many tropical forests. The greatest diversity of reproductive modes is seen in the neotropics, with 21 known from South America, and 14 restricted to tropical regions. One 3 km^2 site in Amazonian Ecuador was found to contain 86 species, which were mainly opportunistic feeders, but had a wide variety of breeding sites. Ten types of egg-laying behaviour were distinguished, which included laying the eggs in water, on vegetation, in tree cavities, in a depression in the soil, in foam-nests on vegetation above water (Figure 2.14a), and on the back of the female (Figure 2.14b).

(a)   (b)

Figure 2.14   (a) A frog foam-nest above water. (b) Tadpoles on the back of a female toad.

Male frogs call to attract females. Usually, male frogs call in the vicinity of only one or two other frog species that are also calling, but at certain times and in certain places, particularly in temporary pools at the start of the rainy season, 10–15 species can be encountered. It is likely that to avoid interference between the calls of males of different species (so that males of each species can be heard by the females of that species), frogs have competed for signal space, which has led to the evolution of adaptations in frequency and timing of the calls.

Tropical forest frogs and toads have also been found to be temporally and spatially segregated, with some species being nocturnal (both tree-dwelling, and ground-dwelling) and other diurnally active (but only ground-dwelling ones).

○ Describe the types of spatial and temporal partitioning that frogs and toads exhibit, and name the temporal and spatial type *not* exhibited by these species in the neotropics.

● The temporal partitioning is day–night, i.e. there are diurnal and nocturnal species. The spatial partitioning is canopy–ground, i.e. there are arboreal and ground-dwelling species. There are no frog or toad species in the neotropics that are both diurnal and arboreal.

2.3 Origins of high species richness and diversity in tropical forests

2.3.1 Introduction

Zoologists have long debated, and still are debating, *why* there are so many species of plants and animals in tropical forests: how did this richness originate and how is it maintained? According to some theories, climax tropical forests are *closed* systems that have reached equilibrium; other theories treat them as *open* systems where the number of species can continue to increase. In some parts of the tropics, the primitive trees and organisms are similar to those found in the fossil record, supporting the view that similar forests existed some 30–50 million years ago. However, the time theory cannot readily be tested and, by itself, cannot explain how the high species richness of tropical forests is maintained.

Although some of the land surfaces on which tropical forests are found are very ancient, such as tropical Africa, others, such as a large part of Amazonia, are more recent in their origin. Also, whereas it was commonly believed that most of the tropics escaped the drastic climatic changes of the glaciations, it is now felt that the alternating dry and wet periods associated with these climate changes probably caused forest areas to contract and expand respectively. This would have had the effect of fragmenting the forest and its animal and plant populations for long periods, before the forest was rejoined. It has been suggested that the contraction of the forest into a number of smaller fragments (known as **refugia**, as they were refuges for the animal and plant populations left) may have encouraged speciation. You may see this referred to as 'the Pleistocene refugia theory'. It is explored further in Section 2.3.2.

Another factor, apart from age and the contraction and expansion of tropical forests, that directly contributes to species richness, is the current climate. Tropical ecosystems illustrate the general rule that in all classes of organism (with a few exceptions) the number of species increases from the poles towards the Equator (Figure 2.15).

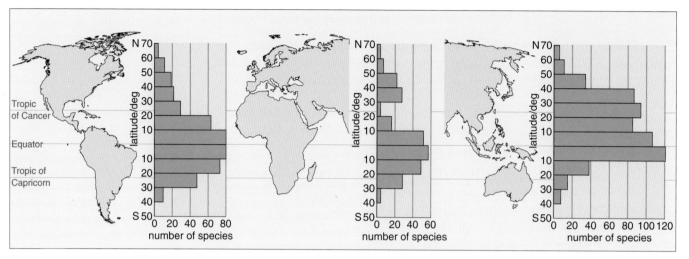

Figure 2.15   Swallowtail butterfly species numbers plotted against latitude: species richness increases towards the Equator.

Within the tropics, plant species diversity is strongly correlated with annual rainfall. Although moisture is important, high mean temperatures are important too, as shown by the higher species richness of lowland forests compared with montane forests that have similar rainfall patterns.

Tropical climates can also produce increased rates of *animal* speciation, compared with the temperate ones, by allowing resources such as flowers, fruits and young leaves to be available throughout the year. This is because tropical climates lack a very cold or very dry spell. The year-long period of resource availability allows a finer division of these resources and ensures more complex plant–animal interactions in the tropics, compared with those in temperate ecosystems.

Survival and extinction rates are very different in the tropics compared with temperate areas. It has been put forward that, whereas the main struggle for survival in the temperate zone is fought against the cold or drought, in the tropics this struggle is between species.

2.3.2 The refugia theory

The origins of the refugia theory come from a paper by Jurgen Haffer published in the journal *Science* in 1969. The author reported that the only way he could explain the amount of diversity and endemism in the bird fauna of tropical America, was if the forest had once consisted of a number of widely separated refugia ((Figure 2.16). Haffer suggested that the most fragmentation took place during the drier conditions of the last glacial period (which began about 100 000 years ago), resulting in a corridor of grassy savanna that separated the forests of the upper and lower Amazon Basin. Similar conclusions have been drawn by other zoologists from the study of South American butterflies, lizards and other animals. Although conclusive evidence from the study of the flora in the region is still lacking, detailed examination of the distribution of trees in several families has found much to support the idea of forest fragmentation about 100 000 years ago. A similar pattern of forest refugia has also been proposed for African tropical forests.

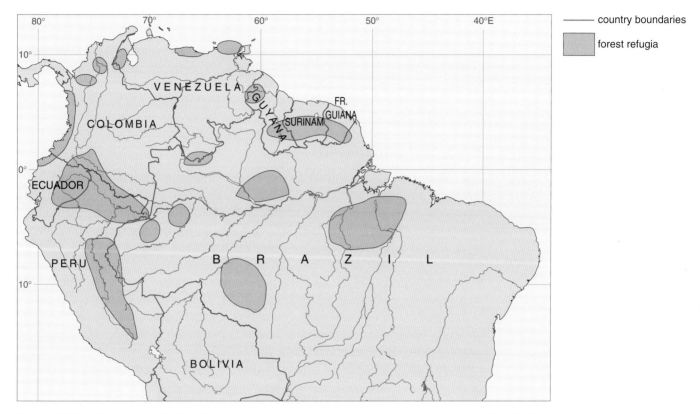

Figure 2.16   The 16 possible locations for forest refugia in tropical America, as proposed by Haffer.

Earlier efforts to use the refugia theory to pinpoint areas of high biological diversity are increasingly disputed. Some of the proposed refugia may have been too cold to support rainforest during glacial periods. Also, the locations of at least some of the proposed refugia may be artefacts of collecting efforts.

Other authors familiar with Amazonia propose an alternative hypothesis to explain current butterfly species distributions. A series of ridges, which were thrown up when the Andes range of mountains was formed, define the boundaries of different species distributions, especially butterflies, across the Amazon region, apparently much better than the refugia theory.

2.4   Tropical forest layers

Forest habitats are often thought of in terms of layers or strata. In a temperate broadleaved forest, such as a European oakwood, these layers are fairly clearly defined (as described in *Life*, p. 158). Commonly, the woody plants form two layers, one of all tall trees and the other of shrubs and smaller trees. Below these are one or more layers of herbs and undershrubs, the ground layers, and sometimes also a layer of mosses and liverworts close to the ground. Each of these layers is composed of a group of plants of similar life-form that play a similar role in the community to which they belong.

In tropical forests, layers are much more numerous and less clearly defined. As in other plant communities, however, the plants in a tropical forest can be grouped into a limited number of groups with a discernable, albeit complicated, three-dimensional distribution.

Here we have attempted to simplify these layers into three generalized strata: soil and leaf-litter, the understorey, and the canopy. We have chosen these three strata purely to provide examples of stratification with respect to both flora and fauna, highlighting the spatial partitioning and specializations brought about by existing in respective strata. Other examples of spatial partitioning will be found elsewhere in this section.

2.4.1 Soil and leaf-litter

The soil surface, and leaf-litter layer on top of it, together constitute the recycling interface; it is here that organic material such as leaves and fruits are broken down, and the nutrients thereby released become available for uptake by the forests' root system (see Section 3.3). In wetter tropical forests, the soil and litter layer seldom desiccates, allowing many life-forms in contact with this microhabitat to exist solely here, and in strictly aquatic or cave habitats. Polychaete worms and the multi-legged *Peripatus* appear to have remained unchanged for millions of years (Figure 2.17). They owe their antiquity, in part, to the steady state of forest floor conditions, which have not applied selective pressures on the species. In certain forests that are periodically inundated with water, making them unsuitable for termites and ants, the niche of litter disposer is filled instead by a variety of often colourful land crabs (Figure 2.18).

Figure 2.17 *Peripatus*, often described as a 'living fossil'. This creature is also called the 'walking worm' or 'velvet worm'.

Figure 2.18 A forest land crab (*Gecarcinus ruricola*).

2.4.2 Understorey

The understorey layers of a tropical forest contrast sharply with the forest canopy, especially in terms of microclimate. The understorey layers are typically dark, the tree crowns efficiently cutting out (by absorption or reflection) 97–99% of the light incident on the canopy and thus preventing it from reaching the lower layers. The light that does penetrate below the canopy is mostly transmitted through leaves or reflected off them; therefore its spectral quality is changed.

○ What part of the spectrum is enriched after sunlight has been filtered by the canopy? (*Hint*: Refer to Figure 3.3 in Block 3, Part 2 *Life*.)

● The infrared; this portion of the spectrum contributes a large proportion of the total radiation penetrating the forest understorey.

Both the temperature and the humidity in the lower layers are much more constant than above the forest canopy, and there is little air movement. All small plants, including the seedlings and juveniles of lianas, must be adapted to these low-illumination conditions, unless they rely on openings in the canopy, as with the shade-intolerant (light-demanding) tree species.

Some epiphyte species start life on the ground and climb up to the canopy using the boles (trunks) of trees. They lay their leaves in a characteristic pattern up the bole of the tree until eventually the epiphyte finds a branch or tree fork to settle on, and then contact with the ground is severed (Figure 2.19).

Figure 2.19   A Trinidadian epiphyte species climbing up a tree bole.

The humid and dark lower layers of the understorey of tropical forests are inhabited by the shade-dwellers. The animals of the understorey tend to be coloured like their surroundings and thus avoid being noticed by predators and prey. This strategy is known as **crypsis** (Figure 2.20a and b). It contrasts with the adoption of bright warning coloration, which warns potential predators of an animal's toxicity (Figure 2.20c and d). The tropical forest jaguars of Brazil are frequently darker than those that inhabit more open areas, for example. Forest-floor understorey butterflies, such as the grass-feeding butterflies of the Satyridae family, are generally brown with a few stripes and eye-spots, blending in to their forest-floor background of dead leaf-litter. The owl butterfly (Figure 2.20a) of the *Caligo* genus is also brown and in addition appears to have bark- or wood-like markings, which make it very difficult to see when at rest with its wings closed.

The neotropical forest floor is populated by large rodents that also tend to be brown and occasionally dappled, like the paca (*Agouti paca*, Figure 2.20b), as if imitating sunflecks, which daub the forest understorey. Asian forest floors are not dominated by large rodents, but by miniature ungulates such as the muntjac deer, *Muntiacus muntjak*, now spreading successfully through the UK after a few individuals escaped from Woburn Safari Park, Bedfordshire.

Figure 2.20   (a) Owl butterfly (*Caligo* sp.), showing cryptic bark-like markings. (b) The paca (*Agouti paca*) of South America, showing dark brown cryptic coloration. (c) Warning coloration of the poisonous arrow tree frog (*Dendrobates truncatus*). (d) Warning coloration of the coral snake (*Micrurius dissoleucus*).

2.4.3 Canopy

The canopy of the tropical forest is a starkly different microclimate to the dark, shaded forest interior and is where primary contact with the elements, chiefly sunshine and rain, occurs. Life in the canopy is adapted to trap as much of the sunshine as possible, but also to cope with desiccation. Tree crowns compete with each other to fill as much space as they can in the canopy, resulting in a patchwork of crowns meshing together to form a canopy blanket (Figure 2.21). Individual tree crowns are only a little, if at all, shaded by their neighbours and their leaves are almost fully exposed to the Sun. They experience the whole range of temperature and humidity of the atmosphere, and air movement around them is unimpeded.

In Guyana, occasional trees of very tall species, such as *Hymenaea courbaril* and *Peltogyne pubescens*, occur as emergent trees that rise above the forest canopy. These taller emergent tree species (Figure 2.22) are rarely found in the main canopy.

○ Describe the competitive advantage emergent trees may have over canopy trees? Can you think of any potential disadvantages?

● Emergent trees gain all the sunlight they need without having to compete with species in the canopy. A disadvantage might be that, being taller, they are more likely to suffer from drought if there are any dry spells, and from damage during strong winds or hurricanes.

On the upper boles, in the crowns, light-demanding epiphytes are present. These plants have various adaptations that enable them to resist desiccation. Their leaves

(a)

(b)

Figure 2.21 (a) Looking up at a forest canopy of *Pentaclethra macroloba* trees in Trinidad. (b) Hemispherical photo of the canopy through a fish-eye camera lens.

are leathery with thick cuticles, and some have water-storage organs, such as some orchid species which have pseudobulbs. As we have seen earlier, the bromeliads abundant in the neotropics have leaf bases arranged to enclose a space or 'tank' that collects water (see Section 2.2.2).

There is roughly $1800\,kg\,km^{-2}$ of mammal biomass in Amazonia, about half of which is composed of arboreal (mainly canopy-dwelling) species. The other half is made up of ground-dwelling species. A similar pattern is found in birds that are large enough to be considered game species, with half of the $100\,kg\,km^{-2}$ of this avian biomass being arboreal and half ground-dwelling. This vertical distribution pattern appears to reflect the availability of food resources. Fruit production in tropical forests is in the range of a few tonnes fresh weight per hectare per year. This is a primary resource base that supports about 80% of the animal biomass of the Amazonian forest. Most of this production comes from the upper strata, with understorey trees and shrubs making a minor contribution. The ground-dwelling birds and mammals also use this food resource once it has fallen to the ground.

The butterfly fauna of tropical forests has also been found to be stratified vertically, with certain Charaxinae species being almost entirely canopy-dwelling, feeding on the fruits found there (Figure 2.23), and others, such as the small Satyridae species, being restricted to the forest floor.

Many primate species are arboreal, very rarely venturing to the ground. These include the red howler and black spider monkeys of the neotropics (Figure 2.24), and many of the gibbon species in Asia. This restriction to the forest canopy has obvious implications for forest fragmentation, where breaks of open ground between fragments of forest may act as barriers to movement.

Figure 2.22 Two emergent trees within a forest canopy.

◄ **Figure 2.23**   *Memphis eribotes*, a neotropical canopy-dwelling butterfly species of the Charaxinae family.

Figure 2.24   Two typically arboreal primate species: (a) the neotropical black spider monkey (*Ateles paniscus*) carrying its young offspring; (b) the red howler monkey (*Alouatta seniculus*).

2.4.4 Gaps and edges

Gaps in the forest are usually the result of a single tree-fall, but can also be caused by bole, branch or crown breakage, the fall of liana tangles or the domino effect of a large tree taking down others. The direct result of a tree-fall is a gap in the forest and an area of affected vegetation. Forests tend to have a natural level of gap formation, and from data collected on four neotropical forests, this has been estimated at 1% of the forest area per year, with an average gap size of about 100 m² per gap. The number and size of light gaps formed in the canopy can be increased by local winds, lightning strikes, landslides and hurricanes.

The initial size of a gap is decisive to the course of subsequent regeneration processes. A small gap may be closed rapidly by the crowns of trees surviving or recovering from the impact of the tree-fall, but a large gap may take longer to recover as it waits to be filled by juvenile trees. Gaps in the canopy are important for the regeneration of many rainforest species and represent a different micro-environment compared to the forest understorey. Not only is light quantity enhanced in gaps, but there are also differences in light quality, soil and air temperature, nutrient and water availability in the soil, and the water saturation deficit of the air. The events following a tree-fall until final replacement by a new canopy tree has been termed 'gap-phase regeneration'.

The sunlit arena created on the forest floor by a new canopy gap is the setting for a 'race' to the canopy. Sunlight on the forest floor is a prerequisite for germination and establishment of the shade-intolerant (light-demanding) tree species. The most successful group in this early stage of succession are the rapid-growing **pioneer** species characteristic of young secondary forest. As time goes on, most of the short-lived light-demanding species are replaced by more slow-growing shade-tolerant species characteristic of mature forest.

Edges also offer an abrupt contrast in microclimate to that of the forest interior. Many true forest-dwelling species appear to shy away from forest edges, and it has been shown that the effects of a forest edge can penetrate as far as 200–250 m into the forest.

○ What is the largest forest fragment size (assume it is circular) in hectares that has no core forest unaffected by the edge effect? Assume this edge effect penetrates 200 m. (1 hectare = 100 m × 100 m or 10 000 m² and $\pi = 3.14$.)

● The area of the patch is πr^2, where r is the radius of the patch, in this case 200 m. Thus the patch area = $3.14 \times (200\,\text{m})^2 = 125\,600\,\text{m}^2$. To convert this value to hectares we divide by 10 000, giving 12.6 ha (to 3 significant figures).

2.5 Summary of Section 2

1 Tropical forest covers 6% of the world's land area and contains 50% of its species.

2 Biodiversity refers to the whole array of life-forms and their components (genes, species, habitats and ecosystems).

3 The term 'species diversity' includes a measure of abundance as well as the number of species.

4 Gamma diversity (the number of species in a region) = alpha diversity (the number of species within a habitat) × beta diversity (the number of habitats within a region).

5 Tropical forests can be extremely rich in the number of tree species that they contain, but others can be dominated by one tree species (monodominant).

6 Some characteristics features are found in all tropical forests, such as tree buttresses, cauliflory and entire leaf margins of tree leaves.

7 Coexistence amongst species through specialization enables the partitioning of resources which may be time, space, food or breeding sites.

8 There are many theories as to why tropical forests are so species-rich compared to temperate ones. These include the time factor, the fragmentation and reconnecting of forest refugia in dry and wet times, and current climatic conditions ensuring all-year-round availability of certain resources.

9 Tropical ecosystems illustrate the general rule that in nearly all classes of organism the number of species increases from the poles to the Equator.

10 Crypsis and warning coloration are two contrasting strategies for avoiding predation.

11 The effect of a forest edge, the edge effect, can penetrate 200–250 m into a forest.

Question 2.1

In your own words, distinguish between the terms 'species richness' and 'species diversity'. Use your own made-up example of tropical tree species in one hectare of tropical forest to illustrate your answer. (*A short paragraph*)

Question 2.2

Summarize, in a table, the major differences between the rainforests of the neotropics and those of Malesia.

3 Interactions

We have come across various ways in which tropical forest habitats are partitioned, both temporally and spatially, allowing a vast array of species to coexist. This has resulted in more than half of all the Earth's species, known to date, being found in tropical forests. The tropical forest presents us with a paradox, therefore. On the one hand, there is an incredible specialization, which allows coexistence. On the other hand, the teeming life is in intense competition. Plants fight with each other for light, water, nutrients and space. Animals struggle with each other for available food, mates and territories. Predators hunt for prey incessantly, whilst potential prey employ a wide range of cunning and trickery to avoid being eaten.

In this section, we will explore some of the processes behind the interactions that take place in tropical forests, linking plants with animals in this dynamic functioning ecosystem, and some of the processes that have led to these interactions evolving in the first place.

3.1 Trophic levels

The tropical forest, like all other ecosystems, can be thought of as a complex piece of machinery for harnessing and processing the energy from the Sun and making it available for all other life in the forest to exist and to prosper. As you learnt in Block 3, Part 2 *Life*, it can be explored in terms of the energy that flows through the system. The idea of energy flow in a food chain will be briefly reviewed. As in almost all other ecosystems, energy flow in tropical forests begins with the energy emitted from the Sun and captured by those organisms that can use it to manufacture their own food, by the process of photosynthesis. In the tropical forest system, these organisms are overwhelmingly plants. Those organisms that can produce their own food through photosynthesis are known as primary producers or autotrophs. The producers form the *first trophic level* of the ecosystem's flow of energy, capturing the energy of the Sun and converting this energy to chemical energy in the form of carbohydrates. Animals cannot harness the energy from the Sun directly, and so gain their energy by consuming plants which have done so, or other animals that have fed on plants. The *second trophic level* is made up of those animals that feed exclusively on plants, the herbivores or primary consumers.

○ How does the energy from the Sun reach a leaf-eating beetle?

● Leaves photosynthesize, converting the Sun's energy into chemical energy in the form of carbohydrates. The beetle then eats leaves, assimilating the energy stored in the leaves as carbohydrates.

Animals that feed on herbivores are known as secondary consumers or carnivores and constitute the *third trophic level*. The *fourth trophic level* (where it exists) comprises the carnivores that feed on other carnivores, the tertiary consumers. As opposed to the autotrophs, which can manufacture their own food, those organisms unable to do so are called heterotrophs.

In terms of biomass, the forest can be viewed as a pyramid, made up of the four trophic levels. The primary producers of the first trophic level are the largest in terms of biomass and make up the base of the pyramid. The second level has less biomass than the first, the third less than the second, and the fourth trophic level has the least.

○ Would a mature tropical forest be better described as a storage or a detritus ecosystem?

● When mature, there is no net addition to the energy stored in the system's biomass. Instead, the huge annual net production is consumed and respired by a combination of herbivory and decomposition. The system is therefore better described as a detritus ecosystem.

3.1.1 Top predators

The predator that is not prey to any other animal in the system, at the top of the food web pyramid, is known as the **top predator**. In many cases this is a large carnivore, but as we shall see later, this is not always the case.

In South America, several big cats are the top predators in tropical forest habitats. The largest of these is the jaguar (*Panthera onca*, Figure 3.1) which is found today in most of both Central and South America, although its original range is thought to have shrunk by 50%. More than 85 prey species have been identified in the jaguar's diet, including large herbivores such as tapirs, deer and peccaries, although a jaguar will eat almost anything it can catch — its diet reflecting the relative populations of mammal prey species. When studying secretive creatures such as the jaguar, that also are relatively low in abundance, the study of individual animals can help build up a picture of the species as a whole. With jaguars, and many other vertebrates in tropical forest and other ecosystems, this has been achieved through **radio-telemetry**. This technique involves attaching a radio-transmitter device to the individual animal in question, and then following the transmissions using a receiver, for as long as the message can be received. In the case of jaguars and big cats generally, this process is achieved by trapping and sedating an individual, attaching a radio-collar while it is sedated, and then releasing the animal once it has come round. Several animals can be tracked at once by using different radio frequencies for different individuals. Although the jaguar has been characterized as being primarily nocturnal, radio-telemetry work has revealed that these animals are often active during the daytime, with activity peaks around dawn and dusk. Jaguars have been found to be active for 50–60% of each 24-hour period. This work also helped to reveal that the mean daily distance travelled by an individual is significantly greater for males (3.3 km) than for females (1.8 km). Jaguars of both sexes tended to travel further each day during the dry season. Another study found that radio-collared male jaguars tended to remain within small areas (average 2.5 km^2) for a week at a time before shifting in a single night to other parts of their range.

Similar radio-telemetry work on the smaller neotropical forest cat, the ocelot (*Felis pardalis*, Figure 3.2), has shown that these animals are strongly nocturnal, resting in trees or dense bush in the daytime. Ocelots are generally

Figure 3.1 The jaguar (*Panthera onca*) is the top predator (secondary or tertiary consumer) in most of the tropical forests of the Americas.

active for more than half of each 24-hour period and the mean daily travel distances range from 1.8–7.6 km, with males travelling up to twice as far as females. Other studies have found ground-dwelling and nocturnal rodents to be the mainstay of the ocelot's diet. The most frequently taken prey species are those of relatively high abundance, and include mice, rats, opossums and armadillos. Ocelots will also take larger prey, including lesser anteaters, red brocket deer, squirrel monkeys and land tortoises (the legs of a very young tortoise have been found in an ocelot's stomach). However, most prey taken weighs less than 1–3% of an ocelot's body weight. Ocelots also vary their hunting behaviour to take advantage of seasonal changes in prey abundance, such as spawning fish and land crabs in the wet season. The results of two studies on the ocelot's diet, based on scat (faeces) analysis, are shown in Table 3.1.

Figure 3.2 An ocelot (*Felix pardalis*) that has been fitted with a radio-collar, prior to release back into the wild (Trinidad).

Table 3.1 Results of two separate studies on the composition of an ocelot's diet.

Study 1	Study 2
65% small rodents	66% small mammals
18% reptiles (mostly iguanas)	12% reptiles
7% crustaceans and fish	5% bats and tree-dwelling mammals
6% medium-sized mammals	5% large rodents
4% birds	10% birds
	2% fish

Some tropical forest systems, especially on small islands, do not have some of the top predators associated with mainland forests, such as the big cats. On the Caribbean island of Puerto Rico, for example, where big cats are absent, the top predator is a whip scorpion (*Phrynus longipes*, Figure 3.3).

○   What effect might you expect to find in lower trophic levels, in terms of biomass, in tropical forest systems where top carnivore predators are absent?

●   A greater biomass of those predators' prey items such as small mammals.

Figure 3.3 A whip scorpion (*Phrynus longipes*). The whip scorpion is more closely related to spiders than to true scorpions. It is not poisonous, but because it mimics true scorpions, it appears more dangerous than it really is.

3.2 Food chains and webs

As we have already seen, interactions between floral and faunal components of the tropical forest, as with all ecosystems, can be represented in terms of how energy is transferred between trophic levels of the forest community — from the producers, which harness the Sun's energy, through herbivores and carnivores, to the top predator. A food chain is a representation of who eats what or whom. For a particular part or the whole of an ecosystem, a more intricate and complete map, linking many food chains, is a food *web*. Every animal or plant has a place in a food chain and food web. For example, a simple food chain might consist of an ocelot as the top predator, which eats agoutis (small mammals), which feed on nuts and roots:

Brazil nut $\longrightarrow$ agouti $\longrightarrow$ ocelot

More links could be added to this food chain, to include other plant materials in the agouti's diet and other animals that are the ocelot's prey, and also what they feed on, so building up a food web.

○ What species is at the top of the Teign Valley food web described in Section 3.2.4 of Block 3, Part 2 *Life*?

● The sparrowhawk (*Accipiter*).

An intriguing example of an unusual food chain that has evolved in tropical forests is described below.

3.2.1 Army ants, antbirds and antbutterflies

Tropical forests provide many unusual examples of links in food webs. One such story surrounds an ant species that raids for prey items in swarms or columns, thus earning the name of army ant. One very conspicuous army ant species in the neotropics is *Eciton burchelli*, whose colony usually comprises a single queen and about 400 000 soldiers and workers (Figure 3.4a). Each colony lives to an approximate 35-day rhythm, with 21 days spent raiding out from a central bivouac or nest site, and the other 14 moving to a new site. A raid can consist of 200 000 ants moving along a front 15 m wide(!), capturing and eating roughly 40 g dry weight of animal matter per day, of which half are prey ants and their broods. It takes the prey ants 100 days to recover to half their original abundance. An ant raid can reduce the abundance of crickets and cockroaches in the leaf-litter by half, but immigrants replace these individuals in a week.

Many species of birds, primarily in the antbird family, Formicariidae (Figure 3.4b), follow these swarms of army ants and feed on the insects flushed from the leaf-litter by the ants.

Females of antbutterfly species (Ithominae subfamily, Figure 3.4c) are attracted to these ant swarms also, and feed on the antbird droppings. The presence of antbirds at ant swarms therefore provides a regular and predictable supply of nutrients for these butterfly species. Reproduction in most kinds of butterflies is thought to be limited by the amount of nitrogenous reserves accumulated during larval feeding, which are needed for egg production. Most temperate butterfly

(a)

(b)

(c)

Figure 3.4   (a) A swarm of army ants (*Eciton burchelli*). (b) A warbling antbird (*Huypocnemis cantator*). (c) An antbutterfly (*Tithoria* sp.)

species feed exclusively on flower nectar, obtaining sugar for metabolic energy, but little nitrogen; for these species, larval feeding may be the only source of nitrogenous reserves. Therefore, the reproductive potential of these species (i.e. the number of eggs produced in the lifetime of the females) is thought to be directly determined by the amount of nutrients accumulated during larval feeding. Antbutterflies feeding on bird droppings are thought to be using partially digested proteins as a source of protein for egg production. Some antbutterfly species are known to live at least four months, and can produce egg clusters throughout their lifetime.

Another fascinating feature of this system is that butterfly species that attempt to profit from this predictable source of nitrogenous compounds are open to attack from the antbirds that are feeding on insects flushed out by the ant swarm. However, antbutterflies are not attacked by the insectivorous antbirds at the swarm, due to their bright orange, yellow and black warning colours (Figure 3.4c), which alert potential predators to their poisonous nature. Some of these butterflies are poisonous because they contain toxic substances that were collected by the caterpillars (larvae) when they were feeding on their specific host-plant of the nightshade family. Other antbutterfly species are brightly coloured, mimicking the poisonous species, but they are not actually poisonous themselves. This difference is explained in Box 3.1.

Box 3.1   *Mimicry*

When an owl butterfly (*Caligo* spp.) is resting on a tree trunk or feeding on rotting fruits on the forest floor, it looks very similar to dead leaves. It is cryptic, and may pass unnoticed by an insectivorous bird. This is its protective resemblance (commonly known as camouflage). If the butterfly moves and is spotted and eaten by the bird, the bird may look slightly harder at other dead leaves. If the bird attacked and ate a tiger butterfly (*Tithorea harmonia*), however, it would react very differently. It would regurgitate its crop, fluff up its feathers and wipe its bill on its perch, generally indicating it had just eaten something very disagreeable. If another tiger butterfly were to fly past, the bird would probably ignore it. The bird has learnt to associate the colour pattern with a nasty taste, and thus the butterfly has gained some protection subsequently through mimicry. This protective **mimicry** can be divided into two classes: Batesian mimicry, where a palatable species mimics the colour pattern of an unpalatable species and thus gains protection by *falsely* advertising unpalatability; and Müllerian mimicry, where several unpalatable model species share the same colour pattern, thereby reinforcing each other's protection because predators learn more rapidly that the colour is a warning.

3.2.2 Farming Brazil nuts

As described so graphically by Wallace (Section 2.1.1), many tropical tree species in mixed forest occur singly, with nearest neighbours of the same species occurring sometimes hundreds of metres away. Harvesting forest tree products thus becomes an arduous test of endurance and fitness. So why not grow a monoculture of a tropical crop species, as we do with so many other crops?

Available data suggest that population interaction is more significant in regulating the structure and function of communities in the humid tropics than it is in temperate regions. The roles of bats, birds, insects and other animals that feed on seeds in open forest clearings are examples of this. The Brazil nut tree (*Bertholletia excelsa*) depends on certain bees in the family Meliponidae for pollination, and its seeds have to pass through the digestive tracts of certain rodents to be able to germinate. When the Brazil nut tree is not in flower, the bees depend on the flowers of other small trees. If these other small trees and sources of flower nectar are not included in the plantation, then the Brazil nut trees will not be fertilized and no Brazil nuts will be produced. This mechanism was learnt about the hard way, after a plantation of Brazil nut trees had been established over a number of years, and not a single Brazil nut was harvested!

3.2.3 Food webs

As we have seen, tropical forests are extremely species-rich. As we have also seen, insects are a dominant group of the world's flora and fauna (57% of all described plant and animal species are insects). Put these two ideas together, and you get an inkling of the problems faced by scientists who are working in tropical forests on insect groups that are largely undescribed. Progress in estimating insect diversity, and in understanding insect community dynamics, can be made by building local inventories of species diversity, and studying the trophic structure of particular communities. An example of one such insect community is that composed of leaf miners and their predators. Leaf miners are insect species that, as larvae, feed on the layer of leaf tissue between the upper and lower epidermis, leaving a trail as they feed, which is usually characteristic for each leaf-mining species (the work of the holly leaf miner is commonly seen in UK hedgerows). These leaf miners then pupate and emerge as adults.

In the food web shown in Figure 3.5, which is from Costa Rica, 92 species of leaf miner were found. Their major predators are another group of insects called parasitoids, so named because they live within a host, like a parasite. But unlike most parasites, parasitoids invariably kill the host, so they are in fact predators. In this case, a total of 93 species of parasitoid were found, all but one of which were from the wasp order, Hymenoptera (which also includes the bees and ants); they parasitize the leaf miners by laying their eggs inside them. The food web in Figure 3.5 shows the association between 86 species of parasitoid and 63 species of leaf miner. Some parasitoid species were found to parasitize single leaf miner species, others to parasitize several different leaf miners. This example illustrates the complexity of food webs at this very small scale (the community of leaf mining insects on leaves, and their parasitoids) in one locality in Costa Rica, and highlights the difficulties therefore faced by entomologists in describing not only these insect communities in all tropical forests, but all the other insect communities too.

parasitoid species

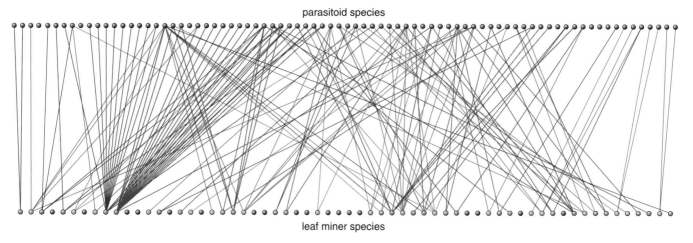

leaf miner species

Figure 3.5   The parasitoid food web for leaf miners in a tropical dry forest in Costa Rica.

3.3   Nutrient cycling

Most, but not all, tropical forest soils are low in nutrients, and hence the main features of nutrient cycling in tropical forests is to conserve these nutrients within the system and recycle them efficiently. In general, the tropical forest nutrient cycle is similar to that of temperate forests, but there are some distinct tropical features. One of these is the great quantity of the biomass of mature tropical forests and the relatively high proportion of the total stock of nutrients that is held in aerial parts such as leaves and reproductive tissues, as opposed to in the soil. This may be a safeguard against losses of nutrients by leaching (washing away by rain) and soil erosion. Another distinctive feature of tropical forest nutrient cycles is that they are virtually leak-proof. Nutrients leave the aerial biomass when leaves are lost and become leaf-litter. Most of these nutrients reach the forest floor rapidly, although some material gets trapped in epiphytes, or in forks between branches to form aerial soil.

The nutrients released from decaying animals, leaves and wood, usually do not move directly to the soil or roots of trees. Instead they pass through a whole series of smaller-scale cycles within the organic portion of the soil. Soil detritivore arthropods chew the leaves and break down the complex organic compounds into simpler ones, which are more readily available for other soil organisms. Decomposition of leaves and wood can also begin with invasion of the tissue by bacteria and fungi. As enzymes released from the fungal hyphae continue to break down the complex organic compounds, so the nutrient concentration in the litter is increased, favouring bacterial colonization. Bacterial activity alters the nature of the substrate, rendering it more favourable to colonization by fungi, and so the process continues.

○ How does the distribution of nutrient reserves in a tropical forest differ from that in a temperate one?

● A much higher proportion of the nutrients are held by living tissue in tropical forests, while more reside in the soil of temperate forests.

The relative digestibility of litter is influenced in part by the carbon-to-nitrogen ratio and the content of phenolic compounds. A low C : N ratio usually indicates a high concentration of all nutrients, and consequently a more favourable environment for decomposers. A low phenolic content makes the litter more favourable to attack by detritivores such as woodlice, millipedes, fly larvae, springtails and a whole variety of worms. Tropical trees tend to produce relatively digestible litter compared to that of other forests.

Many of the released nutrients are readily reabsorbed back into trees through dense networks of fine roots, which tend to be situated near the soil surface. Uptake by the roots of many trees is enhanced by the presence of symbiotic fungi which grow on the surface and within the cortex of the roots.

○ What is the scientific name for such fungi?

● Mycorrhizas. (The role of mycorrhizas in nutrient uptake was described in Block 3, Part 2 *Life*.)

○ Why is the litter layer on the floor of tropical forests often thin in spite of high productivity in the canopy?

● The litter is readily digestible by detritivores and decomposers and is therefore rapidly consumed once it has fallen to the ground, preventing the build-up of a thick litter layer.

3.4 Summary of Section 3

1 Tropical forest ecosystems can be explored, like all ecosystems, in terms of how energy flows through the system, through different trophic levels.

2 In terms of biomass, the forest can be seen as a pyramid, with the first trophic level forming the base and the fourth trophic level the top.

3 Radio-telemetry is a technique that allows individual animals to be tracked and elements of their behaviour, such as the size of their territory and daily distance travelled, elucidated.

4 Protective mimicry can be divided into two classes: Batesian mimicry, where there is an unpalatable species and a palatable mimic; and Müllerian mimicry, where several unpalatable model species share the same colour pattern.

5 Food chains and webs are intricate and often poorly understood in tropical forests. The farming of Brazil nuts is a classic example where omitting one link in the chain has had disastrous consequences.

6 Nutrient cycling in tropical forests is extremely efficient, and so avoids losses from soils which are often low in nutrients; mycorrhizal fungi play an important role.

4

The future of tropical forests

Under the present climate of non-sustainable forest exploitation and clearance, the future of tropical forests is uncertain and precarious. Tropical forests are disappearing rapidly, a process that is not readily reversible. Increased anthropogenic (human-induced) pressure on tropical forests results not only in loss of habitat, but also in a clear reduction in the quality of the remaining habitat. It has been estimated that 92.1×10^6 ha of closed forest were lost between 1980 and 1990. Of this, 40% (36.6×10^6 ha) went to 'other land cover', which includes permanent agriculture, cattle ranching and water reservoirs, and represents a complete loss of cover and woody biomass; 26% (24.1×10^6 ha) went into shrubs and short-term agriculture, which represents deforestation, but with some woody biomass remaining; 20% (18.2×10^6 ha) went into open forest or longer-term agriculture, representing degradation involving the loss of approximately half the biomass of the original forest cover; 10% (9.2×10^6 ha) went to fragmented forest, representing partial deforestation, with a loss of approximately two-thirds of the original forest cover; and 4% (4×10^6 ha) went to forestry plantations.

This section will cover topics pertaining to the future of the Earth's tropical forests, including some of the effects that need to be taken into account when discussing tropical forest conservation, tropical timber exploitation and alternative solutions to preserving tropical forest and all the diversity it contains.

4.1   Forest fragmentation

The reduction in area of tropical forest cover, and a deterioration in the habitat quality of the remaining forest, are not the only deleterious consequences of increased anthropogenic pressures on tropical forests. The remaining forest will also become more and more fragmented into pockets of smaller size, reducing the area of core forest and increasing the ratio of edge forest to core forest. This is important, as the microclimate of the forest core and the forest edge are very different (see Section 2.4.4), with the microclimate of the forest interior being much more stable than outside the forest (both above the canopy and on the outer edge of the forest). The forest interior is much less influenced by changes in wind speed and fluctuations in temperature and humidity. This has a direct effect on plant and animal distributions, with some species being restricted to the forest core and actively avoiding edge and open habitat, and vice versa. As the forest is reduced into fragments (see Figure 4.1), forest core species will be left isolated in smaller patches (compared to the original large block of forest) with smaller population sizes. If these species avoid edges and gaps, then these smaller populations are isolated from other populations. Small populations are often extremely susceptible to sudden, chance reductions in population size (caused by processes such as disease outbreak), which means that they run a much greater risk of becoming extinct. These sudden and chance events are collectively known as **stochastic** processes. The reduction in population size may also contribute to the loss of genetic variation, which in turn can result in impaired reproductive success (see Section 4.3). Hence fragmentation will often result in lower species richness and may reduce the breeding success of some species that persist in fragments.

Figure 4.1 Aerial photo of the Atlantic coastal forest in southeast Brazil, which is extremely fragmented.

One project that is trying to elucidate the effects of fragmentation over a longer time scale than has been done previously, is the Biological Dynamics of Forest Fragment Project (BDFFP) in Manaus, Brazil (Figure 4.2). This project has been studying the effects on many taxonomic groups, of forest fragmentation into 1, 10, 100 and 1000 ha tropical forest fragments. The objective was to determine the critical size of forest ecosystem that should be left after trees were removed from some areas, and to study edge effects. The experimental cutting was finished in 1990, and studies are ongoing. For butterflies it has been estimated that the probable extent of an edge effect is 200–250 m from the edge into the forest. (This was the value quoted in the discussion of edge effects in Section 2.4.4.)

Figure 4.2 The Biological Dynamics of Forest Fragment Project (BDFFP) as seen from the air.

○ Assuming an edge effect of 200 m, how much core forest, i.e. forest unaffected by the influence of the forest edge, would you expect to find in a *square* 100 ha forest fragment? (Recall that 1 ha = 10 000 m².)

● A square forest of area 100 ha covers $100 \times 10\,000\,\text{m}^2 = 10^6\,\text{m}^2$ and therefore has a side of

$$\sqrt{10^6\ \text{m}^2} = 10^3\ \text{m}\ (1000\ \text{m})$$

The outer 200 m is affected by edge effects, leaving an inner core of

$$600\,\text{m} \times 600\,\text{m} = 360\,000\,\text{m}^2 = 36\,\text{ha}$$

i.e. a little more than one-third of the area is unaffected by the edge effect.

Once forest has been fragmented, it is vital to ensure as much connectivity as possible, or at least reduce the amount of inhospitable terrain between fragments. A break of as little as 50 m can have serious effects on species' movements. Implementing measures to increase the connectivity of the fragmented network would seem a logical step in helping to conserve species that are otherwise trapped in small, isolated fragments. The use of **corridors** as conduits to animal and plant movements has received a lot of attention by biologists in the last few decades. Corridors may be effective if they are the same habitat type as that of the fragments they are connecting (but growing a natural tropical forest corridor between two forest fragments would be difficult, for example), and they are of sufficient size to negate some of the edge effects (assuming an edge effect can have influence over at least 200 m, then a corridor greater than 400 m in width would be required).

4.2   Reforestation

Once tropical forest has been removed, there are several options, depending on the aims of the restorers. If the aims are to preserve the biodiversity of the area in its entirety, with the same complement of species and overall general structure, then the forest will have to be allowed to regenerate naturally. Old secondary forest can eventually become indistinguishable from true primary forest, but only if left for a period that is likely to be 200 years or more. Most reforestation projects are aimed at achieving reforestation on a time-scale of much less than 200 years!

Tree cover, even if it does consist of different species to those found in the natural forest it is replacing, will have some benefits, such as helping to prevent flooding and silting, which often result from forest clearance. An extensive planting programme in India's Damodar Valley has reduced floods in catchments above dams, decreased sedimentation and increased water supply for agriculture and drinking.

When reforestation is carried out, it is usually with tree species that have a commercial value, are fast-growing and can be harvested easily. One such species is teak, *Tectona grandis*. Although these monocultural stands are an improvement on bare ground, they do have drawbacks, as they support very little of the flora and fauna found in the natural forest (Figure 4.3).

Figure 4.3   Inside a teak (*Tectona grandis*) plantation forest, showing little understorey growth of other species.

4.3 Conservation

One of the fundamental questions that arises in the tropical forest debate is how best to conserve an area of forest. Traditionally, conservation was seen as preservation of the natural habitat and its people in its entirety, with no development allowed. Nowadays, however, conservation encompasses **sustainable development**, which involves protecting both the habitat and the people while also allowing a level of development such as logging, which can be sustained into the future with minimum damage to the forest or its people. (Biological conservation is discussed in more detail in Topic 11.)

We often think of the number of floral and faunal species that will gain protection when an area of tropical forest is conserved. As stated earlier, the number of species in an area is known as its species richness, whereas the term 'species diversity' incorporates a measure of abundance as well as number of species (Section 2.1). Another facet of biodiversity that we must be aware of when considering conservation issues is **genetic diversity** — the genetic variation that exists within a species. Population genetics is concerned with the genetic variation that exists within and between populations of a given species, from which we can infer the degree of gene exchange between populations, and how isolated (or connected) separated populations are. Generally speaking, a population with plenty of genetic variation will be more likely to have the appropriate genes to adapt and therefore survive if conditions suddenly change, compared to a population with little genetic variation. Therefore, maintaining the species diversity of an area is not the only goal of conservation; maintenance of the genetic diversity of each species is important too.

4.4 Sustainable management

Sustainable forest management usually involves selective logging, where the commercially valuable trees are removed but the rest of the forest and its trees are left as intact as possible. In general, logging practices make such a selective logging operation very difficult to guarantee. The International Tropical Timber Organization (ITTO) is a strong advocate of selective logging, but estimates that sustainable tropical timber operations cover only one-eighth of 1% of tropical forest land.

Trinidad-and-Tobago is one of the few countries in the world where a sustainable-yield selection system of logging takes place. The Periodic Block system, as it is known, has a cutting cycle of 30 years. Thus a forest block is opened for selective logging, and after logging it is closed for a period of 30 years to allow recovery. This logging system is based on a set of tree selection procedures that include not removing all the straightest, most well-formed boles, and leaving a relatively high numbers of the larger fruiting trees, such as the hogplum tree (*Spondias mombin*). Although the hogplum is an economically important tree species, it also has a high wildlife value (many forest animals feed on hogplums). In the past, logs were extracted using oxen (see Figure 4.4), thereby limiting the amount of damage to other trees in the extraction process (although tractors are now being used with increasing frequency).

Figure 4.4 An ox waiting to drag felled logs out of the Periodic Block system forest in Trinidad.

○ Imagine a typical tropical forest made up of tall and straight trees and others that are bent and crooked. What do you think would be the effect on the forest as a whole, with respect to logging in the future, if only the tall, straight trees were removed?

● The forest gene pool would slowly lose the genes for tall, straight boles, and future logging for the commercially valuable straight trees would cease. It would not be sustainable.

4.5   Gap analysis

Gap analysis is a conservation tool for analysing species' distributions. It plots the area of habitat occupied by the target species along with an estimate of the species' potential habitat, i.e. areas that fall within its fundamental niche.

○ Using your knowledge of Block 3, Part 2 *Life*, define the term 'fundamental niche'.

● The range of environments in which a given species could survive in the absence of competition from other species.

On this map, a second layer of information is lain, which shows the extent of areas currently protected or managed for nature conservation. A comparison of the two maps allows *gaps* in the species' protection to be highlighted.

Gap analysis is based on three data sets in the form of mapped distributions of the vegetation (for example tropical forest), protected areas (forest reserves) and the distribution of the species of interest (the white-faced capuchin monkey, *Cebus capucinus*, in the example shown in Figure 4.5). By overlaying these maps, the degree to which a species' distribution is represented in the existing system of protected areas, can be determined. A 'gap' is recognized when a species of conservation importance lacks a sustainable population within a protected area. Once it is known with reasonable precision where the gaps in protection occur, priorities for the next set of conservation actions can be pursued. The ultimate product of a gap analysis is not a list of gaps, but a map showing areas that are a priority for conservation action.

The objectives of a gap analysis may include the following:

• Identifying which species and natural communities are unprotected or inadequately protected on existing areas that are managed primarily for their biodiversity.

• Identifying areas in each region that contain the largest number of unprotected or underprotected species or natural communities. These areas are candidates for change in management status to ensure that all species and natural communities are adequately represented in the network of areas managed primarily for biodiversity.

• Identifying which other areas, in addition to those already identified, must be managed for their biodiversity to ensure the persistence of unprotected or underprotected species or natural communities.

Recognition of the fact that conserving a species in the long term requires maintenance of its genetic diversity has led to conservation initiatives being designed at a regional rather than a local scale. Gap analysis is one step in this design process.

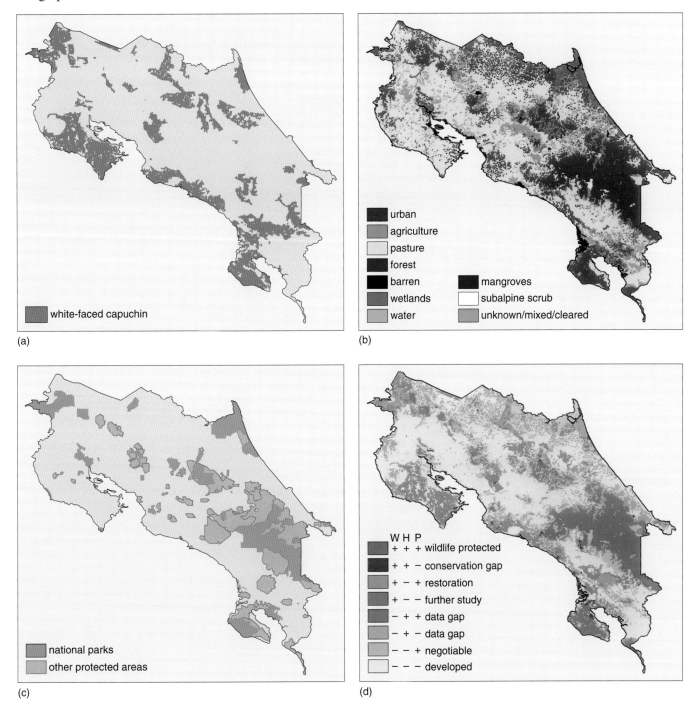

Figure 4.5 A series of maps of the Central American country of Costa Rica, illustrating the process of gap analysis for the rare primate species, the white-faced capuchin monkey (*Cebus capucinus*). (a) The capuchin's range within Costa Rica. (b) The distribution of different habitat types and land uses. The capuchin is confined to forest or forest remnants interspersed in pasture. (c) Areas that have some form of protected status for wildlife. (d) A map showing the product of gap analysis. (W = wildlife species present; H = habitat available; P = protected area.)

4.6   Summary of Section 4

1   Tropical forest loss is a major problem, and remaining forest areas are likely to become fragmented, with an increased edge to core ratio (supporting more edge species and less forest core species).

2   Populations in fragments will be smaller and become more isolated from other fragments and therefore populations, increasing the risk of extinction from chance stochastic events.

3   Connectivity between fragments may be increased by using habitat corridors.

4   Reforestation is preferable to leaving clear-cut areas barren. Time-scales of 200 years are needed for natural regeneration to replace the forest lost. In practice, commercially valuable fast-growing species are the norm. This does have ecological drawbacks, however.

5   Conservation of tropical forest is aimed at preserving the habitat and its people, whereas sustainable management of forest aims to achieve this *and* exploit the forest at low, sustainable levels.

6   Sustainable forest management is usually by means of selective logging practices.

7   As well as conserving species diversity within tropical forests (and other ecosystems), it is important to consider conserving populations' genetic diversity.

8   Gap analysis aims to find the 'gaps' in species and habitat protection so that measures can be taken to offer further protection where necessary.

Question 4.1

Assuming an edge effect of 200 m, what area of core forest, unaffected by the influence of the forest edge, would you expect to find in a *circular* 100 ha forest fragment?

Question 4.2

Using the value you worked out in Question 4.1 and doing similar calculations for other circular fragment sizes, what can you say about the validity of comparing results obtained in tropical forest projects using 1, 10, 100 and 1000 ha fragments?

Learning outcomes for Topic 10

After working through this topic you should be able to:

1   Describe the extent and location of the tropical forest biome and contrast the major tropical forest regions. (*Questions 1.1 and 2.2*)

2   Comment on the use of satellite imagery to monitor the extent of tropical forests. (*Activity 1.1*)

3   List reasons why tropical forests form the most diverse biome on Earth.

4   Distinguish between 'species richness' and 'species diversity' and describe the relationship between alpha, beta and gamma diversity and the geographical scales to which they relate. (*Question 2.1; Activity 2.1*)

5   Explain how the large number of species within a tropical forest are able to coexist.

6   Describe the effect of fragmenting a stand of tropical forest.

7   Contrast the food webs of tropical forests with those of other biomes.

8   Explain why animals use mimicry, and describe two different types of mimicry observed within tropical forests.

9   Compare the cycling and storage of nutrients within tropical forests with those within other biomes.

10 Explain how forest fragmentation affects habitat conditions within the remaining fragments. (*Questions 4.1 and 4.2*)

11 Describe how a tropical forest can be sustainably managed for timber production.

12 Explain why the conservation of genetic diversity within the tropical forest biome is important.

Comments on activities

Activity 1.1

In the first square, in the west (004, 63), you can detect major rivers, minor river tributaries, cloud cover and cloud shadows, and it appears that the area is completely covered by forest. In the second two squares, in the central area (228, 63 and 227, 63), you can make out a major river and to the south of the river, but near to it, a herring-bone of disturbances, which are possibly major and minor logging roads. It appears that the logging operations are using the river to transport the logs. The last two squares (222, 63 and 221, 63) show much less forest and lots of lines and pink areas, which are probably development from suburbs and on into a major city. Some areas are a much lighter green, which may be secondary forest.

Whereas distinguishing between forest and grassland is relatively easy, it is more difficult to discriminate between old-growth forest and regenerating secondary-growth forest. One of the largest sources of error in measuring forest cover originates from this inherent difficulty in distinguishing between similar land cover types.

Activity 2.1

The herpetofauna species list includes a total of 43 species belonging to the Hylidae family of frogs. This value is the gamma diversity, as it is the species richness of the whole of Guyana.

Answers to questions

Question 1.1

Similarities have been found in both the floral composition and the overall structure of the forests:

In terms of floral composition, most of the 36 genera of trees found in the floral fossils of Alaska are found today in Malesia.

Structural similarities include: leaf drip-tips; entire leaf margins (about two-thirds of all leaves); trees predominantly evergreen; and lianas comprising a quarter of all the forest plant species.

Question 2.1

Species richness refers solely to the number of tropical tree species in the one hectare of forest, whereas species diversity includes a measure of the number of individuals of each species. A higher species diversity is found when there is greater evenness or spread of the individuals between each species. In the following example, both plots have the *same* species richness (five), but different species diversities. In the first hectare of forest, there are five species with ten individuals each, so this plot has a high species diversity. Of the five species in the second hectare of forest, one species has 30 individuals, one species has 17 individuals and the remaining three species have one individual each. Therefore in this plot, two species dominate, so there is a lower species diversity.

Question 2.2

Table 2.4 lists the major differences between the rainforests of the neotropics and those of Malesia.

Table 2.4 Answer to Question 2.2.

	Malesia	Neotropics
Total number of species	25 000	85 000
Diptocarp tree species	many	few
Species in Brazil nut family	very few	many
Conifer tree species	many	only one

Question 4.1

100 ha = 1 000 000 m^2. First of all we need to know the radius of the circular patch, which is calculated using the formula for the area of a circle, πr^2, and rearranging it:

$$r = \sqrt{\frac{\text{area}}{\pi}} = \sqrt{\frac{1\,000\,000 \text{ m}^2}{3.14}} = 564.3 \text{ m}$$

The edge effect is 200 m, therefore the radius of the inner core forest, which is not affected by the influence of the edge, is 564.3 m − 200 m = 364.3 m. We use this value of the radius in the formula πr^2 to find the area of the inner core:

$$\text{area} = 3.14 \times (364.3 \text{ m})^2 = 416\,723 \text{ m}^2, \text{ or } 41.7 \text{ ha}$$

The total area of the forest patch is 100 ha; thus only 41.7% of it is core forest.

Question 4.2

As for the 100 ha circular patch in Question 4.1, the radii of the 1, 10 and 1000 ha patches are calculated using the formula

$$r = \sqrt{\frac{\text{area}}{\pi}}$$

The respective radii are

$$\sqrt{\frac{10\,000\,\text{m}^2}{3.14}} = 56.4\,\text{m}$$

$$\sqrt{\frac{100\,000\,\text{m}^2}{3.14}} = 178.5\,\text{m}$$

and

$$\sqrt{\frac{10\,000\,000\,\text{m}^2}{3.14}} = 1785\,\text{m}$$

Thus both 1 ha and 10 ha circular forest fragments are entirely influenced by the edge effect. (In fact, we calculated in Section 2.4.4 that circular fragments up to 12.6 ha have no core forest.)

From the answer to Question 4.1, we know that nearly 60% of a 100 ha patch is influenced by the edge effect.

In a 1000 ha patch, the core region has a radius of (1785 − 200) m = 1585 m, so the area of the core is

$$3.14 \times (1585\,\text{m})^2 = 7\,888\,387\,\text{m}^2,$$
or approximately 789 ha

Thus about 211 ha of a 1000 ha patch, or 21.1%, is affected by the influence of the edge.

Therefore, the results obtained from experiments in 1, 10, 100 and 1000 ha forest fragments cannot be attributed solely to the size difference between fragments. Every result will be due in part to the relative influence of the edge effect on each forest fragment size.

Acknowledgements for Topic 10
Tropical Forests

Grateful acknowledgement is made to the following sources for permission to reproduce material in this book:

Figures 1.1a, 2.3, 2.4, 2.10, 211, 2.14a, 2.19, 2.21, 2.23, 3.2, 3.4c, 4.4: Byron Wood; *Figures 1.1b, 2.14b*: Dr Morely Read/Science Photo Library; *Figures 1.1c, 2.7*: Mike Gillman/Open University; *Figures 2.1, 2.5, 2.20a*: Mike Dodd/Open University; *Figure 2.2*: Michael and Patricia Fogden; *Figure 2.8*: Science Photo Library; *Figure 2.12a*: Robert Tyrell/Oxford Scientific Films; *Figure 2.12b*: Michael Fogden/Oxford Scientific Films; *Figures 2.13a and b*: David Haring/Oxford Scientific Films; *Figures 2.13c, 2.20b*: Partridge Production Limited/Oxford Scientific Films; *Figure 2.17*: P. J. De Vries/Oxford Scientific Films; *Figure 2.18*: E./R. Degginger/Science Photo Library; *Figure 2.20c*: John Wilkinson/Open University; *Figure 2.20d*: George Bernard/Science Photo Library; *Figure 2.21b*: Inga Spence/Holt Studios; *Figure 2.22*: Oxford Scientific Films; *Figure 2.24a*: Art Wolfe/Science Photo Library; *Figure 2.24b*: Aldo Brando/Oxford Scientific Films; *Figure 3.1*: Mike Powles/Oxford Scientific Films; *Figure 3.3*: Courtesy of Joe Wood, copyright © 2000, New Mexico Pest Management; *Figure 3.4a*: Oxford Scientific Films; *Figure 3.3b*: Pattie Murray/AA/Oxford Scientific Films; *Figure 4.1*: Copyright © Michael Giannechini; *Figure 4.2*: The Biological Dynamics of Forest Fragments Project (National Institute for Research in the Amazon and Smithsonian Institution); *Figure 4.3*: Mark Edwards/Stills Pictures; *Figure 4.5*: Savitsky, B. G. and Lacher Jr., T. E. (eds), (1998) 'Wildlife and habitat data collection and analysis', *GIS Methodologies for Developing Conservation Strategies*, Columbia University Press.

Every effort has been made to trace all the copyright owners, but if any has been inadvertently overlooked, the publishers will be pleased to make the necessary arrangements at the first opportunity.

TOPIC 11

BIOLOGICAL CONSERVATION

Mike Gillman

1	**The themes of biological conservation**	**120**
1.1	Introduction	120
1.2	The ecological relationship between species, habitat and climate	123
1.3	Habitat loss and species extinction	124
1.4	Loss of species due to hunting and harvesting	128
1.5	Monitoring the losses (and increases)	131
1.6	Protection and management	132
1.7	Summary of Section 1	134
2	**Monitoring of endangered species**	**135**
2.1	Describing the spatial distribution of a species	135
2.2	Global biodiversity hotspots	137
2.3	Collecting data on population size	140
2.4	Using data on population size	150
2.5	Summary of Section 2	154
3	**Management of habitats**	**155**
3.1	Habitat management and secondary succession	155
3.2	Management of early successional habitats	156
3.3	The relationship between managed and natural woodland	162
3.4	Summary of Section 3	167
	Learning outcomes for Topic 11	**167**
	Acknowledgements	**168**

1   The themes of biological conservation

1.1   Introduction

Biological conservation is a globally important subject, in which science plays a key role in identifying problems and seeking solutions. This topic will build on earlier knowledge from the course, especially from Block 3, Part 2 *Life*. The variety, complexity and international nature of conservation *problems and responses* can be seen in the headlines in Figure 1.1.

As the range of examples in Figure 1.1 implies, biological conservation is a subject that is difficult to delimit or define. The underlying ethos of biological conservation is a desire to avoid (or at least, reduce) the loss of species and habitats. As this section will show, species loss and habitat loss are inextricably linked. Defining the problem and devising an appropriate response requires precise information, which is why *monitoring* the distribution and abundance of species is a vital element in modern biological conservation, as Section 2 demonstrates. Conservation involves much more than a simple preservation of the status quo: Section 3 illustrates that it requires active interference, more properly termed *management*. In summary, this topic is concerned with how science can help identify and quantify conservation problems and suggest possible responses. The purpose of this section is to introduce the themes of biological conservation, and thereby provide a broad definition of the subject.

You should read through Box 1.1 to check you are familiar with some key terms that were introduced in *Life* and will be used throughout the current topic.

Morley fights to keep ban on whaling

Paul Brown
Environment correspondent

Fisheries minister Elliott Morley flew to Japan last night to try to prevent the 17 year ban on commercial whaling being overturned at the International Whaling Commission.
Several small countries which are recipients of

Crime unit to fight trade in rare animals

By Charles Clover
Environment Editor

Ministers are considering tougher sentences for people who trade in endangered species, Michael Meacher, the environment minister, said yesterday at the launch of a national police unit dedicated to wildlife crime. The National Wildlife Crime Intelligence Unit has been set

Police Officers (Acpo), the Home Office and the Scottish Executive. Mr Meacher disclosed at the launch that the Government was about to launch a consultation on the adequacy of the laws on wildlife that the unit would be enforcing.
Under the Control of Trade in Endangered Species regulations of 1997, a maximum sentence of two years is attached to

Coral bleaching is worse than ever in the Great Barrier Reef

White mischief

Jonathan Walter

Record high temperatures have caused unprecedented bleaching of the world's largest coral reef system and led to a stark warning from an international group of marine scientists that the world's tropical coral reefs could die out in 30 years as water temperatures rise due to climate change.

Scottish forests spread

There are more trees in Scotland than at any time since Robert the Bruce's day in the 14th century, the Forestry Commission said yesterday.
The area of woodland has nearly trebled in 100 years, with trees covering more than 17 per cent of the land area. Commercial conifers comprise nearly half of the total.

Allan Wilson, the forestry minister, said it was a "wonderful success story", with forestry supporting more than 10,000 jobs.
But Clifton Bain, of the Royal Society for the Protection of Birds, said: " Many of the forests were planted without taking account of their impact on wildlife."

Senate blocks oil drilling in Alaska's wilderness

By David Rennie in Washington

A CENTRAL plank of President Bush's energy policy was blocked by the Senate yesterday when it rejected a move to allow oil drilling in an Alaskan wilderness.
Allowing oil exploitation in the Arctic national wildlife refuge was at the heart of an energy plan released by Mr Bush a year ago which argued that America needed to produce more of its own oil.
Ari Fleischer, White House press secretary, condemned the vote and said Mr Bush would continue

'Worst ever' GM crop invasion

Figure 1.1 Conservation headlines from national UK newspapers in April and May 2002.

Box 1.1 Defining species, populations and habitat

You should be aware by now that various observable differences allow organisms to be put into named categories, such as elephant, giraffe or oak tree. On closer inspection these categories can be seen to be divisible into a series of smaller categories; for example, there are at least two types of elephant (the African and the Asian elephant, Figure 1.2) distinguishable by their body size (the African is larger) and various other characteristics, such as their teeth and the size of their ears. African elephants can be further divided into three separate types, based on size and habitat preferences: the forest or round-eared; the bush or large-eared and the pygmy.

That two individuals possess different characteristics does not necessarily mean that they belong to different species. A species is defined strictly as comprising all organisms that are able to interbreed and produce viable (healthy and fertile) offspring. Two closely related species may be able to produce hybrids, but these should be infertile. Where individuals *appear* to comprise different species, based on certain characteristics, but in fact are able to interbreed, they may be referred to as subspecies. These subspecies may, over time, evolve to become separate species.

You will recall from *Life* (Section 2.2.4) that it is conventional to give species a latinized scientific name. This is in addition to the organism's common name. The name 'oak' is a common name, derived from the Anglo-Saxon 'ac' meaning fruit or acorn. The common name, the pedunculate oak, is in fact quite a good one, because it identifies an important characteristic of this species (the peduncle, Figure 1.3). Unfortunately, it has a second common name — the common oak! Conversely, one common name may correspond to several species. Therefore, to avoid confusion and ensure that each species has one name that is used internationally, the common or pedunculate oak also has the scientific name *Quercus robur* (Figure 1.4).

Incidentally, the origins of the two parts of this scientific name are quite obscure. *Quercus* is an old Latin name that is used for all oak species. It is possibly associated with the Greek for pig, *choiros,* because pigs are fond of acorns! The species name *robur* apparently means 'strength' or possibly 'hardwood' or 'elite', all suitable names for the oak.

A population is a group of individuals of the same species, all of which have the potential to interbreed (*Life*, Section 2.1) and that live close to each other and are therefore likely to mate with each other. Individuals from different populations of the same species are less likely to interbreed.

A habitat is defined as the type of environment where individuals of a species live (*Life*, Section 2.3.3). These habitats may occur in saltwater, freshwater or on land (terrestrial). This topic deals mainly with terrestrial habitats.

(a)

(b)

Figure 1.2 (a) African and (b) Asian elephants.

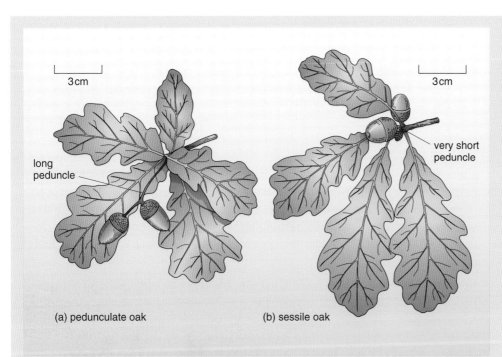

Figure 1.3 (a) Pedunculate oak (*Quercus robur*) and (b) sessile oak (*Quercus petraea*), showing the difference in peduncle length.

Figure 1.4 Pedunculate oak (*Quercus robur*), the common oak species of lowland Britain.

1.2 The ecological relationship between species, habitat and climate

The relationship between climate and broad habitat type (defined with respect to the dominant vegetation) is sufficiently robust to allow predictions of the types of world habitat in relation to rainfall and temperature (*Life*, Figure 2.40). These climatically determined broad habitat types are referred to as biomes, several of which should now be familiar to you from earlier parts of the course (e.g. tropical rainforest, tundra and desert). Each biome contains a characteristic set of species, each of which will usually be restricted to one or more areas in that biome.

○ What natural factors may lead to disruption of the simple scheme of climate determining vegetation type (biome)?

● Animals may have an effect on the vegetation, e.g. through grazing or defaecation. Climate may also have a direct effect on animal distribution. The biome categories could also be divided according to different soil types.

Even within soil types, smaller divisions of habitat may be recognized. For example, individuals of a tree species (which may be one of many tree species in a woodland habitat) can provide the habitat for one or more species of herbivorous insect. Similarly, that herbivorous insect may in turn comprise part of the habitat for one or more species of parasitic insect living in, or on it. Thus we can define scales of habitat, from large climatically influenced biomes, to (micro-) habitats generated by the activities and/or presence of the individuals of particular species.

In Sections 1.3 and 1.4 we consider examples of the conservation problems facing habitats and species.

1.3   Habitat loss and species extinction

Species extinction means the complete (global) loss of a species.

○ What factors may be responsible for the extinction of a species?

● The factors may be divided into *indirect* and *direct* factors reducing the abundance of a species. Indirect factors include loss of habitat, which may affect food source and possible mutualistic species, or degradation of the habitat, e.g. by pollution. Direct factors impacting on species abundance include hunting and disease.

Consideration of the effects on habitat can be divided into the *conditions* and *resources* associated with the habitat (e.g. *Life*, Section 2.4.1). Loss or degradation of the habitat may reduce the quality and quantity of the resources and lead to reduction in quality (appropriateness) of conditions.

In this section we will consider the relationship between **habitat loss** and species extinction. More often we consider the phenomenon of **population extinction**. Species extinction and population extinction are clearly related in that loss of populations, through direct or indirect means, reduces the overall abundance of the species and may contribute to a loss of genetic variation.

1.3.1 Temperate woodland in Britain

The high levels of tropical deforestation since the 1960s are quite rightly raising concerns about the rate of species extinction (Topic 10 *Tropical Forests*, Section 4). However, past levels of habitat destruction in Britain and other temperate regions have been enormous. In 1830, the historian Thomas Carlyle wrote:

> Whoever was up-rooting a thistle, or bramble, or draining a bog, or building himself a house, that man was writing the history of England.

In these terms, the history of lowland England was almost entirely written by the 14th century, when most of the woodland (wildwood) that covered the area following the end of the last glaciation *(c.* 8000 BC; see Figure 1.5) was removed (Figure 1.5). Indeed, it is likely that no wildwood remained in England after the Norman Conquest. In contrast, small areas of Scotland may still retain elements of the original pine and oak woodlands. The removal of the woodland in England began in earnest with the arrival of Neolithic farmers *(c.* 3000 BC). There is evidence that earlier Mesolithic peoples *(c.* 8000–4000 BC), who were primarily hunters, cleared some woodland, possibly by means of fire, to increase the open areas in which they could hunt their prey. Humans were not the only agents of change, however. Elm, which was widespread in Europe prior to 4000 BC, declined dramatically after that period, probably as a result of Dutch elm disease spread by bark beetles.

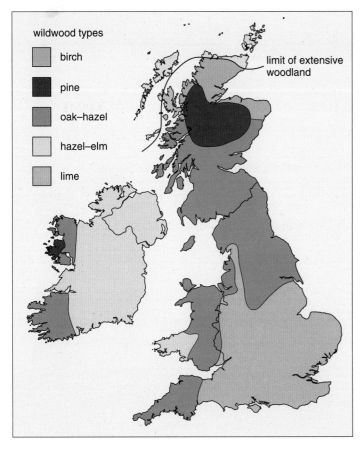

Figure 1.5 Distribution of wildwood types in Britain *c.* 4500 BC. Note that the lime is small-leaved lime (*Tilia cordata*).

In lowland England, the process of deforestation continued after the Romano-British period, with less accessible and less easily worked areas of woodland being cleared. The height of the woodland clearance appears to have been during the 13th and 14th centuries. Along with the woodland went the wolf, wild boar, bear and beaver, partly due to hunting and partly due to habitat loss. Other species have fallen in abundance or become highly fragmented, existing in pockets of 'ancient' woodland, usually more than 500 years old. (Habitat fragmentation was introduced in *Life*, Section 4.1.6, and is revisited in *Tropical Forests* Section 4.1.) A few large tracts of woodland avoided clearance often owing to their protection as, for example, hunting parks. The remaining fragments of ancient woodland can be recognized by some of their resident species that are believed to be characteristic of such habitats, including plants such as wood anemone (*Anemone nemorosa*), oxlip (*Primula elatior*) and small-leaved lime (introduced in *Life,* Figures 2.26 and 4.7) and various animal species, e.g. slugs (Figure 1.6). These species are rarely found in hedgerows or more recent woodland.

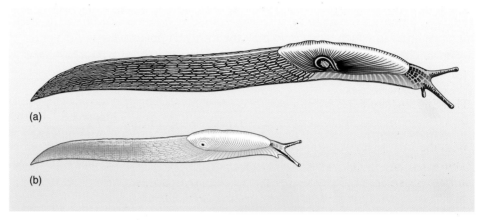

Figure 1.6   Animals as ancient woodland indicators; the slugs (a) *Limax cineroniger* and (b) *Limax tenellus.*

A summary of the changes in woodland in Britain over the last five millennia is provided in Table 1.1.

Table 1.1   Change in area and composition of woodland in Britain over five millennia.

Year	Total woodland area (% of land area)	Area of primary or ancient woodland (% of land area)	Area of coniferous woodland (% of all woodland)
3000 BC	85	85	15
1086 AD	15	5–10	20
1895 AD	4	2	25
1992 AD	11	1.5	70

While the data for 1895 and 1992 are fairly accurate, the data for 1086 and 3000 BC must be viewed with caution. The 1086 data are based on interpretations of, and extrapolations from, the Domesday Book record. The data for 3000 BC can be regarded as the wildwood state (Figure 1.5) before the activities of Neolithic farmers.

○ How do you think the data for 3000 BC were derived?

● Pollen records (see Section 1.5) combined with considerations of landscape, for example, that woodland would have been absent from very high or very wet areas.

Let us concentrate on the more recent data.

○ There appears to have been an *increase* in the total woodland area since 1895. Based on Table 1.1, what is the major contribution to this increase?

● Coniferous woodland (largely due to plantation), which has increased from 25% to 70% of woodland. (Recall the headline on spreading Scottish forests in Figure 1.1.)

The change in the percentage of land area taken up by coniferous woodland (as opposed to the percentage of land area taken up by woodland) can be calculated as follows:

1895   25% of 4%   =   (25/100) × (4/100)   =   1% of land area

1992   70% of 11%   =   (70/100) × (11/100)   =   7.7% of land area

The difference of 6.7% in land area of coniferous woodland therefore explains most of the increase in total woodland area (from 4 to 11%, i.e. 7%).

With the loss of the woodland in lowland England and elsewhere, new semi-natural habitats began to replace the former extensive woodland cover. Brandon, in his assessment of the history of the Sussex landscape, wrote:

> It seems therefore that almost all of the open country in Sussex has been created by man in the process of progressive deforestation which in England led to the virtual destruction of deciduous woodland.
>
> Brandon, P. (1974) *The Sussex Landscape*, Hodder and Stoughton.

In lowland England, heaths developed on the sandy soils, while the chalk and limestone soils became covered with areas of grassland. Figure 1.7 shows the distribution of two plant species in Sussex, the cross-leaved heath (*Erica tetralix*) and the horseshoe vetch (*Hippocrepis comosa*) (*Life* Figure 2.50), which are representative of species restricted to sandy and chalky soils respectively.

○ Recall from *Life* Section 2.4.3, the names of the two plant types represented by *Erica tetralix* and *Hippocrepis comosa*.

● Calcifuge, restricted to acidic soil, and calcicole, restricted to basic soils.

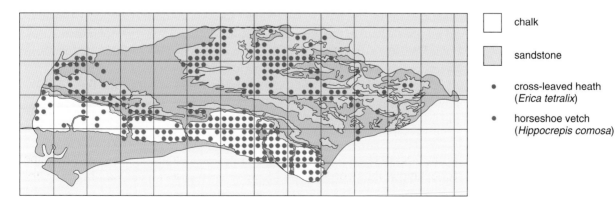

chalk

sandstone

● cross-leaved heath (*Erica tetralix*)

● horseshoe vetch (*Hippocrepis comosa*)

Figure 1.7   Distribution of cross-leaved heath (*Erica tetralix*) and horseshoe vetch (*Hippocrepis comosa*) in Sussex overlaid on chalk and sandstone.

Ironically these two habitats — heathlands and chalk grasslands — are now themselves rare, endangered by human activities. For example, much of the chalk grassland was ploughed up and 'improved' for agriculture with artificial fertilizers during the mid- and late 20th century. This in turn jeopardized the survival of plant and animal species associated with chalk grassland. These species include the early gentian, *Gentianella anglica,* one of the few species restricted to Britain and not also found in mainland Europe.

Problems of habitat loss are exacerbated in species with naturally small areas of occurence, such as species restricted to one or a few small islands. Such species are said to be **endemic** to those islands (or given land area). Many of these island habitats and species have become prominent in the international conservation literature and the popular press. 'Dead as a dodo' is a familiar expression. The dodo was a large, ground-nesting, flightless bird restricted to the island of Mauritius. The bird was last seen in 1680. The Galapagos Islands, the avian inhabitants of which were one of the inspirations for Charles Darwin's theory of natural selection, are home to a number of species found nowhere else. Within 200 years of Captain Cook landing on the islands of Hawaii, more than 20 species of bird had been lost. Such island habitats are often faced with a range of problems, which continue to this day, including tourism and development (Figure 1.8), over-harvesting and introduction of exotic (alien) species such as rats or mongoose.

Figure 1.8   Pressure of tourism: British Virgin Islands surrounded by cruise ships and yachts.

1.4   Loss of species due to hunting and harvesting

Even where a habitat remains intact, species may be reduced in abundance by hunting or harvesting. One of the most poignant examples of this is the passenger pigeon (*Ectopistes migratorius*). This North American species was believed to have been one of the most numerous birds in the world. It was so abundant that during migration it darkened the sky for hours on end. Estimates of bird numbers were in the thousands of millions. In 1808, the naturalist Alexander Wilson estimated a flock to be in excess of 2000 million. By 1870, as Europeans colonized more of North America, the passenger pigeon was identified as a pest and 1000 million birds were killed at one of the last communal nesting sites in Michigan. By 1899 the bird was extinct in the wild and the last passenger pigeon died in Cincinnati Zoo in 1914 (Figure 1.9a).

The koala (*Phascolarctos cinereus*) provides another example of the impact of hunting, with fortunately not the same fate as the passenger pigeon. The koala is the model 'cuddly teddy bear' (although it is not a bear!). In spite of this, millions of these animals were shot in the 1920s for their fur (Figure 1.9b), resulting in the near extinction of the species. In a six-month open season in the state of Queensland (in northeastern Australia), one million koalas were killed. In 1924, two million skins were exported from Australia, leaving the species

Figure 1.9 Victims of hunting: (a) passenger pigeon (photo of last passenger pigeon, Martha); (b) koala (photo of trappers and skins); (c) brown bear and (d) wolf.

extinct in South Australia and close to extinction in New South Wales and Victoria. Partly as a result of public pressure, the koala was declared a protected species by the late 1930s. Unfortunately the koala's habitat, *Eucalyptus* forest, was not protected at the same time. The koala feeds only on *Eucalyptus* trees. Habitat clearance for farmland and urban areas, combined with the effects of drought, disease and bushfires, have produced a series of small and fragmented populations. Estimates of koala population size in 2002 are: 25 000–50 000 in Queensland; 10 000–15 000 in New South Wales, and 10 000–15 000 in South Australia and Victoria. Thus, the upper estimate of the total number of koala individuals remaining in the whole of Australia today, is less than one-tenth of the number that were slaughtered in Queensland in 1927.

In Europe, the demise of the brown bear (*Ursus arctos*, Figure 1.9c) and wolf (*Canis lupus*, Figure 1.9d) was largely due to hunting, trapping and poisoning. Large predators have often been persecuted due to their actual or perceived threat to livestock and game species. Both the brown bear and wolf were formerly widespread throughout Europe, including the islands of Britain (both species) and Ireland (wolf). By 1800, wolves had been eradicated from the British Isles and the coastal lowlands of France, Belgium, Denmark, Germany and Poland. In the following 150 years the reduced continental distribution became increasingly fragmented. By 1973 large populations remained only in Eastern Europe and smaller and isolated populations in former Yugoslavia, Greece, Italy and on the Iberian Peninsula (Figure 1.10). The brown bear's decline began earlier than that of the wolf. It was lost from Denmark 3500 years ago, in Britain by the Middle Ages and in the German lowlands by 1600. There is no doubt that these two species have also suffered from loss and fragmentation of their natural forest habitat, becoming restricted to the more inaccessible areas of Europe. More recently, conservation efforts have resulted in recoveries of both species. For example, the number of brown bears has increased in Russia, the Carpathian Mountains and Scandinavia. Both species are re-colonizing the Alps.

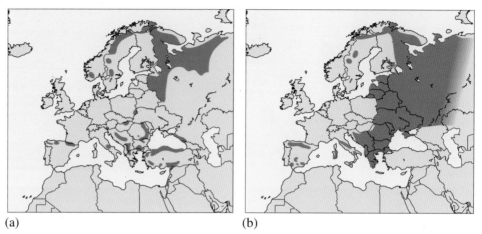

(a)   (b)

Figure 1.10   Current European distribution of (a) brown bear and (b) wolf.

In the tropics there have been marked reductions of species due to hunting, even before the effects of habitat loss (detailed in *Tropical Forests*). The following quotation from John Terborgh, relating to his work in the tropical rainforests of South America (Figure 1.11), is a powerful statement of how a tropical forest habitat may remain intact but its inhabitants lost:

> This study [of primates] would never have taken place were it not for our 'discovery' of Cocha Cashu in 1973. At that time I had been travelling in Peru and other South American countries for ten years, but had never before seen a place in the lowlands that was utterly pristine. While this may seem surprising in view of the vast expanse of primary lowland forest that remains in Amazonia to this day [early 1980s], the fact is that nearly all of it has been exploited in one way or another, for game, furs, natural rubber, Brazil nuts, prime timber etc. Animal populations are especially vulnerable to the slightest encroachment, and appear to melt away before seemingly insignificant human populations. This is because of the devastating effectiveness of modern firearms. A skilled hunter equipped with a standard one-shot 16-gauge shotgun can single-handedly eliminate large birds and mammals within a radius of several hours walk from his dwelling. So, even if the forest looks lush and intact from the air, the appearance is deceptive, much of it is a hollow shell so far as animals are concerned.
>
> Having only seen places which were exposed to hunting, I was not fully aware of this. It was the contrast of Cocha Cashu that put all my previous experience in perspective. Animals are actually plentiful, especially primates. Never before had I seen so many monkeys of so many species, much less monkeys that did not flee at the first hint of a human being.
>
> Terborgh, J. (1983) *Five New World Primates*, Princeton, USA.

In this quotation Terborgh comments on a range of products that can be harvested from tropical forest species. Many valued biological resources can be obtained only by killing individuals, which in turn can contribute to the demise of the species. Large-bodied animal and plant species are especially vulnerable to overharvesting for several reasons. Compared with smaller species, they have a higher yield of product relative to the effort of hunting. They may be more conspicuous and therefore more easily hunted and killed with suitable tools such as rifles or chainsaws. In addition, larger-bodied species generally have lower rates of population increase than smaller-bodied species.

In the remainder of Section 1, we consider the type of scientific response that is appropriate to the problems of species extinction and habitat loss. First, it is necessary to monitor the rates of loss to clarify the extent of the problem. Second, if the outcome of the monitoring points to a problem, conservationists need to consider how the rate of species and habitat loss may be reduced through protection and/or management.

Figure 1.11 Interior of tropical forest in South America.

1.5 Monitoring the losses (and increases)

Monitoring involves recording the abundance of individuals, populations, species or habitats. In particular, monitoring will help to determine *changes* in abundance over time and/or space.

○ Recall examples of monitoring from *Life*.

● Freshwater monitoring, e.g. monitoring of river quality and abundance of freshwater invertebrates by the Environment Agency (*Life* Section 2.4.4).

The technology involved in monitoring varies widely from satellite systems (e.g. the Landsat system described in *Tropical Forests* Section 1.3.2) to the squares of metal (quadrats) used by ecologists to assess plant abundance. These monitoring systems cover spatial scales from thousands of square kilometres to square centimetres (or less for some organisms).

Another dimension — that of time — is often added to the monitoring process. Current species and habitat losses or increases need to be weighed against past changes. However, this is possible only if we have a detailed knowledge of these changes, such as details of the fossil record. Whilst the fossil record is measured over millions of years, the records shown in Figure 1.12 refer to thousands of years. In this technique, cores through soil or lake sediments are taken and the abundance and species identity of pollen in the sediments recorded. The strata from which pollen are taken are then dated to provide a picture of how the vegetation has changed over time. The species composition and extent of the wildwood (Figure 1.5 and Table 1.1), which covered most of western Europe following the last glaciation, is known through evidence of pollen records.

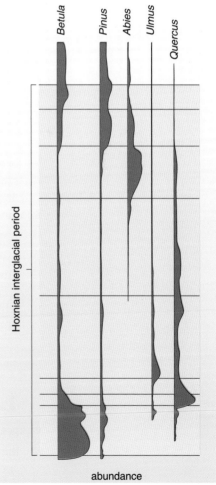

Figure 1.12 Examples of changes in plant species abundance through the Hoxnian interglacial period (400 000 to 380 000 BC) derived from pollen cores.

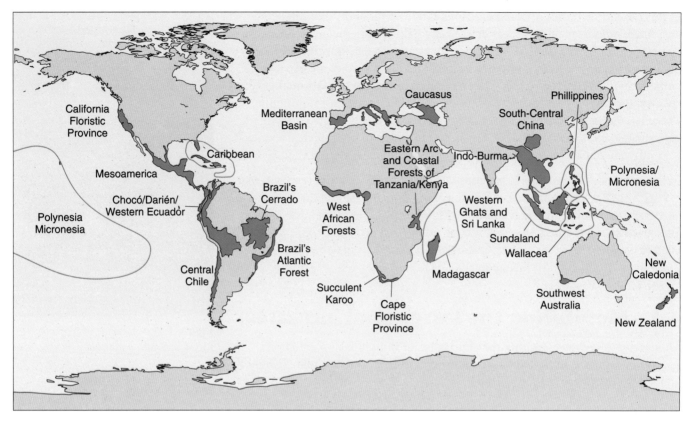

Figure 1.13 Global biodiversity hotspots.

○ Describe how *Betula* species appear to be changing over time.

● They are more abundant at the beginning and end of the interglacial period, suggesting that they prefer relatively cooler climates.

Since the early 1980s there has been a huge increase in monitoring effort and a concomitant increase in the development of databases to accommodate the burgeoning information. One form of synthesis has been the identification of global **biodiversity hotspots**. These are areas of the world with particularly high numbers of endemic species (Figure 1.13). The hotspots encompass a variety of biogeographical regions. Global hotspots are discussed in detail in Section 2.2.

What happens when the monitoring exercises indicate that species or habitat losses are too high? Perhaps a species appears to be rapidly disappearing from a number of locations, or satellite images over several years have recorded a continuing reduction of a particular habitat. Often, the greatest hurdle is not one with which science can assist. Economic or land-use conflicts stand in the way of many effective responses to conservation problems. If we assume that these conflicts can be resolved, then there are two clear options available: protection and/or management.

1.6 Protection and management

Protection, at its crudest, simply involves demarcating an area and preventing the factors that are causing the species or habitat loss (e.g. hunting, deforestation and pollution) from continuing. We have already noted that some woodland was protected for hunting. One of the first examples of an area protected unselfishly for the benefit of its wildlife was Zofinsky Prales (the primeval forest of Zofin), in southern Bohemia, which was designated on 28 August 1838 by its owner Count Jiri Frantisek August Buquoy. The area was established in order to protect the forest birds from exploitation, so that they might 'thrive in peace'. Over 150 years later the reserve is still in existence, with a status of National Preserve (the highest protection category in the Czech Republic), and supports a rich flora and fauna, including 35–40 nesting bird species.

Environmental law is not covered in this topic, but scientific knowledge is required to help in the drafting of environmental legislation, for example, helping to specify the number, size and status of protected areas. An example of the extent of protection of habitat in Britain is given in Table 1.2, with reference to the protection of ancient woodlands. The protection is due to a variety of designations, for example Site of Special Scientific Interest (SSSI), National Nature Reserve (NNR), National Trust (NT) land and other protection.

Table 1.2   Area (hectares) of protected woodland in England, Wales and Scotland.

	England	Wales	Scotland
total semi-natural (not plantation)	205 992	30 658	80 024
semi-natural under protection	44 667	5116	25 553
% protected	22	17	32

Protection of a site is a first step in the process of halting or reversing species or habitat loss, though it is frequently not used in isolation. Coupled with this protection is *management* of the habitat or species (which will often include public awareness and education programmes not considered in this topic).

It often appears contradictory that in order to prevent the demise of one species, a habitat has to be managed in a way that may reduce the abundance of other species! The conservation of rare plants in grassland often requires cutting of scrub and/or grazing to remove existing trees and to prevent the establishment of young trees in the area (Figure 1.14). In essence we need to mimic the 'management' of the area (maintain the deforestation) by our predecessors, who created the habitat in the first place.

Much of the management of terrestrial ecosystems involves controlling the progress of *succession*, i.e. directional vegetation change as described in *Life*. Unless areas of terrestrial habitat are managed, then under many climatic conditions we will be left with little grassland and lots of scrub and eventually, mature woodland. One might argue that this is a good thing. In other places trees are being lost at an alarming rate; furthermore, this is what the original landscape was like before the interference of humans. The answer to this conundrum (although the 'answer' really dodges the issue) is that humans have to make choices — we are able to both destroy and create habitats. We can maintain the grassland through management such as sheep grazing, and so promote

Figure 1.14   An area of scrub and recently cleared areas on chalk grassland at Malling Down near Lewes, East Sussex.

133

the growth of otherwise rare grassland plants. Conversely, we can leave grassland to return to scrub and woodland — we can even accelerate that process by adding tree seeds and excluding natural grazers such as rabbits. These choices need to be made in the light of the long-term history of the site and the conflicts with other land uses, but they are our choices and should be made in an informed and objective way. Hopefully, by the time you reach the end of this topic you will be able to appreciate more clearly the scientific criteria upon which such choices are made. The details and implications of some present-day management procedures will be described in Section 3 and some you have already encountered in *Life* Section 2.4.6, e.g. coppicing.

1.7   Summary of Section 1

1   The primary problems of biological conservation are species extinction (population extinction) and habitat loss.

2   Responses by conservationists to these problems are divided into monitoring (to objectively clarify the extent of the problem), protection and management.

3   Species extinction and habitat loss are linked phenomena — the latter is often a cause of the former, along with direct factors such as hunting. These problems may be especially severe for endemic species, e.g. those restricted to one or a few islands.

4   Monitoring of species abundance and habitat area and quality is undertaken at different spatial scales, using a variety of technologies. The changes that can be identified by these methods occur over a wide range of time-scales. One form of summary of monitoring data is global biodiversity hotspots.

5   When species or habitat losses are identified, the next step is usually to implement protection and/or management policies. Management of terrestrial habitats is often achieved by controlling the natural succession from grassland to woodland.

Monitoring of endangered species

2

The most fundamental and natural division of the world's flora and fauna is into different species. This is also the biological unit that is most often described as in need of conservation, whether it be the giant panda (*Ailuropoda melanoleuca*), the blue whale (*Balaenoptera musculus*) or the lady's slipper orchid (*Cypripedium* sp.) (Figure 2.1). This section will consider monitoring of single species and combinations of species (we will not explicitly describe habitat monitoring). It has been noted that species are divided up into populations and several sections that follow will deal with monitoring the distribution and size of populations. We will also discuss the change in size of one or more populations over time as a prerequisite for determining the extent to which those populations require protection and/or management.

(b)

(a)

(c)

Figure 2.1 Some endangered species: (a) blue whale, (b) giant panda and (c) lady's slipper orchid.

2.1 Describing the spatial distribution of a species

Identification of an endangered species is not a trivial task. Information on some species is so limited that it is extremely difficult to determine if they are threatened: indeed, it is likely that there are many species, particularly in the tropics, whose existence is unknown. On a more positive note, there are now accurate, detailed data available for many plants, vertebrates and some invertebrates, e.g. butterflies and molluscs, which allow rare and possibly endangered species to be identified and their status quantified. This is particularly true in Britain and parts of North America and mainland Europe, where there is a long tradition of studying ecology and natural history and a relatively high ratio of ecologists to species. However, as we shall see in Section 2.2, species information for the rest of the Earth is rapidly catching up.

135

A fundamental set of data for any species (endangered or not) is its spatial distribution. The **geographical range** describes the full (global) limits of distribution of a species (e.g. the geographical range for the green-winged orchid, *Orchis morio* (Figure 2.2), is shown in Figure 2.3). Within those boundaries the species are unlikely to be continually distributed. The determination of the boundaries and the measurement of patchiness of distribution within the boundaries are dependent on the quality of data available. Such data are often in the form of records of presence or absence within grid squares of a certain size. Recall from *Life* (Section 4.3.1) the distribution of butterflies in 10 km × 10 km squares in Britain. We need to be aware that the presence of a species within a given square may be just one individual or a large population. Altering the grid size helps to clarify the situation. For example, Figure 2.3b–d shows the distribution of the green-winged orchid (*Orchis morio*) in England, Wales and Scotland at three different grid sizes. (Note the plant is absent from much of Scotland, Figure 2.3a.) The species was formerly widespread in lowland meadows and was believed to have been one of the commonest species in Europe.

○ What happens to the apparent abundance of the species as the grid size is reduced?

● At larger grid sizes (100 km × 100 km, Figure 2.3b), the plant appears to be relatively common, but it is increasingly rare at finer resolution in the centre of its range in Britain (Buckinghamshire), i.e. it occupies a smaller fraction of the grid squares (Figure 2.2c and d). This is because the distribution is very patchy, which is masked at coarser grid scales. It is likely that the results are similar or worse over many parts of the species' distribution in Britain.

Figure 2.2   Green-winged orchid, *Orchis morio*.

Box 2.1   *The appeal of orchids*

Orchids are examples of plants that have great aesthetic appeal. In Britain there are about 50 species of orchid, including the green-winged orchid (Figure 2.2). Orchid species are often described as the most beautiful and unusual plants and have generated great interest amongst naturalists. The herbalist John Gerard described, in 1597, the various kinds of 'fox-stones' (orchids):

> There be divers kindes of Fox-stones, differing very much in shape of their leaves, as also in floures: some have floures, wherein is to be seen the shape of sundry sorts of living creatures [see *Note* below]; some the shape and proportion of flies, in other gnats, some humble bees, others like unto honey Bees; some like Butter-flies, and others like Waspes that be dead; some yellow of colour, others

white; some purple mixed with red, others of a brown overworne colour ... There is no great use of these in physicke, but they are chiefly regarded for the pleasant and beautifull floures wherewith Nature hath seemed to play and disport her selfe.

Note   One feature of orchids is their unusual methods of pollination. A bizarre and often quoted mechanism, which relates to Gerard's description, is that of pseudo-copulation. In this pollination system the flower is assumed to resemble a small insect such as a bee. These pollinating animals are then attracted to this flower by its colour, shape and smell in the belief that they have located a mate. They attempt to mate with the flower and in so doing collect its pollen. This is a good story; unfortunately, the evidence for pseudo-copulation is rather scant.

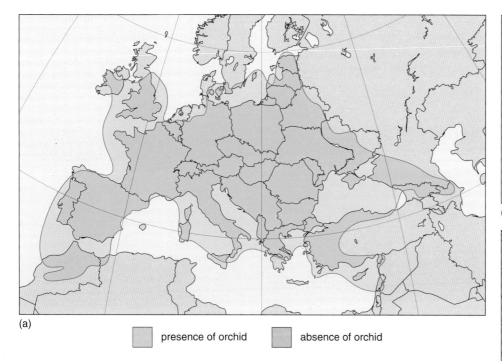

(a)

▢ presence of orchid   ▢ absence of orchid

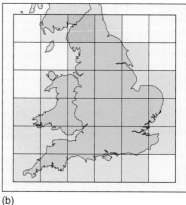

(b)

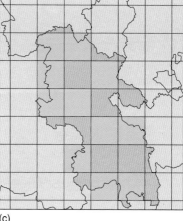

(c)

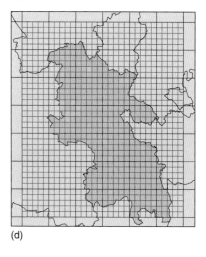

(d)

Data on spatial distribution are useful in two ways when embarking on a detailed population study. First, they indicate whether a species has a limited and/or declining range (if recorded over several years). Second, they can help locate ideal sites for detailed population studies. For example, if the range of a species appears to be contracting, it may be helpful to study a population at the edge of the species range where the factors reducing its range might be expected to be most pronounced. In fact, the choice of site for detailed population studies is often determined by more pragmatic reasons; in particular, studies may need to be undertaken over long periods of time and involve expensive equipment, so site security is important. Sites may also be chosen for ease of access, because the history of the site is well known or because a particular form of management is in place.

○ Recall from Section 1.5 how geographical range data are also used.

● They are used in the compilation of global biodiversity hotspot data.

We will now return to the issue of global biodiversity hotspots and consider the data set in more detail.

2.2   Global biodiversity hotspots

The work of Norman Myers, Russell Mittermeier and collaborators, has resulted in the identification of more than 20 global biodiversity hotspots. This is based on endemism of plant and vertebrate species and the extent of threat to the habitats containing those species. These hotspots cover 44% of the world's plants. Because the data set uses endemic species we can directly relate loss of these habitats to loss of species. The vast data set has been gathered from more than 800 publications. In Tables 2.1–2.5, a summary of the data is given for five

Figure 2.3   (a) Geographical range of green-winged orchid (*Orchis morio*) in Europe. Distribution of *Orchis morio* at three different spatial scales of measurement: (b) 100 km × 100 km in Britain; (c) 10 km × 10 km and (d) 2 km × 2 km in Buckinghamshire.

different taxonomic groups. The data are presented in terms of the number of endemic species and the percentage of all species in the taxonomic group that are endemic.

Table 2.1   Top five global hotspots for *plant* species ranked by number of endemics and percentage of species that are endemic.

Hotspot	Number of endemic species	Hotspot	% of species that are endemic
Tropical Andes	20 000	Tropical Andes	6.7
Sundaland	15 000	Sundaland	5.0
Mediterranean Basin	13 000	Mediterranean Basin	4.3
Madagascar	9700	Madagascar	3.2
Brazil's Atlantic Forest	8000	Brazil's Atlantic Forest	2.7

You will see that the top five plant hotspots are the same regardless of whether we look at number of endemic species or the percentage of species that are endemic. Now look at Tables 2.2–2.5, where the patterns look rather different.

Table 2.2   Top five global hotspots for *bird* species ranked by number of endemics and percentage of species that are endemic.

Hotspot	Number of endemic species	Hotspot	% of species that are endemic
Tropical Andes	677	Polynesia/Micronesia	69
Mesoamerica	251	Madagascar	55
Wallacea	249	Tropical Andes	41
Madagascar	199	Wallacea	36
Philippines	183	Philippines	33

Table 2.3   Top five global hotspots for *mammal* species ranked by number of endemics and percentage of species that are endemic.

Hotspot	Number of endemic species	Hotspot	% of species that are endemic
Mesoamerica	210	New Zealand*	100
Wallacea	123	Madagascar	75
Sundaland	115	New Caledonia*	67
Philippines	111	Wallacea	61
Madagascar	84	Polynesia/Micronesia*	56

* New Zealand has only three mammals; New Caledonia, nine mammals; and Polynesia/Micronesia, sixteen mammals.

Table 2.4 Top five global hotspots for *reptile* species ranked by number of endemics and percentage of species that are endemic.

Hotspot	Number of endemic species	Hotspot	% of species that are endemic
Caribbean	418	New Zealand	100
Mesoamerica	391	Madagascar	92
Madagascar	301	New Caledonia	86
Sundaland	268	Caribbean	84
Tropical Andes	218	Wallacea	65

Table 2.5 Top five global hotspots for *amphibian* species ranked by number of endemics and percentage of species that are endemic

Hotspot	Number of endemic species	Hotspot	% of species that are endemic
Tropical Andes	604	New Zealand	100
Mesoamerica	307	Polynesia	100
Brazil's Atlantic Forest	253	Madagascar	99
Chocó/Darién/Western Ecuador	210	Brazil's Atlantic Forest	90
Sundaland	179	Caribbean	87

○ What geographical factors appear to contribute to high levels of percentage endemism?

● Geographical isolation, usually due to separation by large bodies of water, seems to be critical. Thus the islands of New Zealand, Madagascar, Polynesia and the Caribbean all have high percentage endemism for a range of animal groups.

○ Given this answer, how can the high percentage endemism of the Tropical Andes (plants and birds) and the Atlantic Forest in Brazil (amphibians) be explained?

● These habitats are also geographically isolated, not by water but by other terrestrial habitats. Montane habitats such as the Andes may be separated from one another and other mountain ranges. The Atlantic Forest lies on the coast of Brazil (Figure 1.13) and is separated from the enormous Amazonian forest by savanna grassland.

Geographical isolation interacts with an organism's power of dispersal to generate different mean levels of percentage endemism. For example, the amphibians, with relatively low dispersal ability, have very high values of percentage endemism (the top five range from 100% to 87%). For the amphibians the isolation is more profound than for the birds, with a range from 69% to 33%, many of which are able to cross geographical barriers through flight. The lowest range is for the plants, from 6.7% to 2.7%, an order of magnitude lower than birds, which is due in part to the generally high dispersal of seed and pollen.

A final point is that habitat fragmentation increases the levels of geographical isolation. For example, the Atlantic Forest, geographically separated from Amazonia, is now highly fragmented along its length. The normal assumption is that this high fragmentation will lead to higher rates of population and species extinction in the remaining fragments. However, if the populations persist over sufficient numbers of generations, then the geographical isolation may also lead to reproductive isolation and the emergence of new species.

2.3 Collecting data on population size

2.3.1 A case study of the green-winged orchid, *Orchis morio*

A study undertaken by the author and Mike Dodd of the Open University, will be used to illustrate the monitoring of population size of the green-winged orchid

(*Orchis morio*) introduced in Section 2.1. We have noted that this species is likely to have declined in abundance. This is probably due to changing management of its lowland meadow habitat, for example the increased use of fertilizer, and alterations in grazing or hay-cutting regimes. Certainly the species seems much more restricted in central England (Buckinghamshire) compared to 80 years ago. When collating data for his county flora in the 1920s, Druce stated that green-winged orchids were common throughout the county. Since 1996 we have been studying green-winged orchid distribution and abundance in detail at one of the Buckinghamshire sites at Pilch Field near Great Horwood (Figure 2.4).

Figure 2.4   Ridge and furrow in grassland of Pilch Field.

Figure 2.5   A surveying apparatus, known as a Total Station System, being used to locate samples in a meadow.

Pilch Field is in fact two fields covering a total area of 12 ha (hectares). The site is designated a Site of Special Scientific Interest of neutral grassland, tall fen and scrub, managed by cattle grazing. Green-winged orchids are found across the whole site and were first noted in 1982, although they are likely to have been there earlier. Our study has focused on a 1ha area within a ridge and furrow system in the larger field (recall the ridge and furrow system introduced in *Life* Figure 3.54). The location of every flowering spike within the 1ha plot has been recorded within the first two weeks of May using a Total Station System (Figure 2.5). This is a piece of surveying equipment that bounces an infrared beam off a reflector and calculates the position of the reflector. As a result, individuals can be located with an accuracy of approximately +/−2 cm. This system enables us to map the distribution of the species and to locate the same individuals in successive years.

A summary of the data is provided in Table 2.6. You will see that there are three types of data. The number of flowering spikes is simply the number of stems of the orchid with flowers. However, one individual plant may produce more than one flowering spike. The second row of the table gives an estimate of the number of individuals, referred to as genets (genetic individuals). This is an estimate based on the proximity of flowering spikes — in this case we assumed that all spikes within 3 cm of each other were the same individual. Of course, we cannot be certain about this, unless we dig up the plants (which, apart from being illegal, would defeat the object of the exercise!). An important result of this work, which has been shown in other orchid studies, is that orchids will re-flower in consecutive years. Indeed, the same individual may flower in one year and then miss a year, remaining in a dormant state underground, and then re-flower in the third year. This makes the determination of new genets difficult.

○ Why does dormancy create problems for estimating the number of new genets?

● Because an apparently new genet may be one that has missed one or more years and therefore not been previously recorded in the study.

This is less of a problem as the study increases in duration, because it is likely that orchids will only fail to flower over one or two years. So, as the study continues, we should be getting better estimates of the number of new genets (and the number of true deaths, as opposed to missing individuals).

Table 2.6 Change in total number of green-winged orchid flowering spikes, total number of genets and new genets from 1996 to 2002 within the 1 ha study area in Pilch Field.

Year	1996	1997	1998	1999	2000	2001	2002
number of flowering spikes	121	84	144	92	158	146	182
number of flowering genets	—	71	118	76	129	118	151
new genets	—	55	66	34	39	26	46

The record of locations was incomplete for 1996 as one large clump was omitted (although the total number of spikes was recorded). Since then every flowering spike has been recorded in each year.

The data in Table 2.6 have been plotted in Figure 2.6. This is to help you visualize the patterns of change over time. Although there are relatively few data, so that it is possible to see the patterns from the tabulated data alone, it is always good practice to generate a graph. Note that it is often the case that the population size is given as the number of individuals of a given species living in a particular area, e.g. 1 km^2, and is referred to as the population density. In the present study, the numbers of spikes or genets could be given as the number per square metre (e.g. in 1996 there were $121/10\,000 = 0.0121$ flowering spikes m^{-2}), which can be more easily compared with the results of other studies.

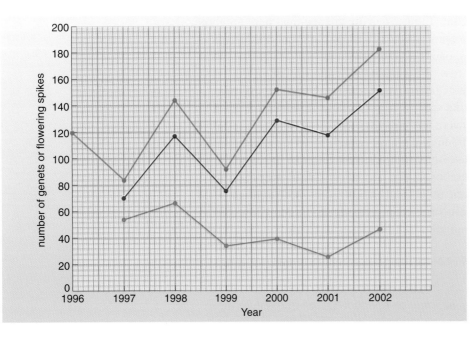

Figure 2.6   The change in number of flowering spikes and genets over time in Pilch Field.

○ What conclusions can you draw from the pattern of change over time in green-winged orchids at Pilch Field?

● It is difficult to determine patterns because this is a relatively short study and there are two relatively low years (1997 and 1999). However, it appears that the population is increasing in size, with the rate of increase being maintained by a high number of new recruits in 2002. Earlier comparisons with new genets are flawed owing to the difficulty of distinguishing new genets from those old genets which missed flowering in 1996 or 1996 and 1997.

This run of data will need to continue for about 20 years before we can draw firm conclusions. However, it is interesting to speculate on the reasons for patterns of change. For example, the low number of flowers in 1999 seems to be due to low levels of re-flowering in that year (i.e. not due to death). Low numbers of flowers were also found at a separate site in south Oxford and therefore may be related to regional climate patterns.

2.3.2 Can all the individuals in a population be counted?

The green-winged orchid example has shown that, even for apparently conspicuous and sessile organisms such as flowering plants, it can be very difficult to count *all* the individuals of a particular species in a study area.

○ Which two problems can you envisage in attempting to count *all* the individuals in a population?

● First, the population may exist over a very large area, so there is the practical problem of effort and time. Second, the individuals may be mobile, very small or hidden (as in the green-winged orchid example). Not even plants sit above ground all year and wait to be counted! Indeed, even in a population with very large immobile individuals, such as the pedunculate oak *(Quercus robur),* there will be small inconspicuous individuals, such as viable acorns in the soil or newly emerged seedlings.

The solution to the first of these problems is to *sample* the population, counting individuals or measuring the population size in some other way in a small area (or set of areas), and then extrapolate to the whole population. Sampling provides an *estimate* of the size of a population in the absence of an absolute count of all the individuals in the whole population. Ecological research has provided conservation workers with a large number of different sampling techniques. Three of these techniques — sampling by direct counting of individuals, indirect methods and using traps — are described in Sections 2.3.3–2.3.5, along with examples of their application.

2.3.3 Sampling by direct counting of individuals

We have already seen an example of direct counting of individual plants with the green-winged orchids. Many plant species cannot be sampled in this way because it is difficult to distinguish between individuals; indeed, this was the case with the orchids, where different flower spikes may be from one or more genetic individuals (genets). For this reason, much plant sampling involves estimates of relative abundance using measures such as percentage cover or biomass (e.g. see *Life* Section 2.3.1).

For many mobile animals, direct observation and recording of individuals is the best method of sampling. In Britain, the **Butterfly Monitoring Scheme** uses a walk-and-count method to assess the size of butterfly populations at different places in the country, covering about 80–100 sites each year since 1976. A recorder follows the same walk, usually once a week, from the beginning of April until the end of September (26 recording weeks), noting the number of adult butterflies of each species seen. The route for the walk is chosen to cover a range of habitats under specified weather conditions during which adult butterflies are known to fly. The recorder walks at a uniform pace recording all butterflies within 5 metres. All but two butterfly species in Britain can be distinguished in flight. This sampling system provides detail beyond that in Section 4.3.1 of *Life*, where the changes in butterfly numbers in 10 km × 10 km squares were described.

The walk-and-count sampling system gives an estimate of the *relative* size of different populations of the same species; for example, there are four times as many individuals in one population compared to another. These comparisons within species can be made between geographical locations and time periods (years). As the study does not provide an estimate of the absolute size of the population, comparisons between species are difficult.

The butterfly monitoring scheme is beginning to produce some intriguing results. The most striking of these is that common species are increasing in both abundance and geographical range; in short, common species are getting commoner. Consider the following six species that have increased their range (Table 2.7).

Table 2.7   Walk-and-count results from butterfly species that have increased their range. The *P* (probability) values indicate the significance of the trend or relationship.

Common name	Scientific name	Larval food plant	Overall increase?	Increase in abundance in east higher than west?
small skipper	*Thymelicus sylvestris*	grasses	yes ($P < 0.05$)	no
large skipper	*Ochlodes venata*	grasses	no significant overall change	yes ($P < 0.001$)
peacock	*Inachis io*	stinging nettle	yes ($P < 0.05$)	yes ($P < 0.001$)
comma	*Polygonia c-album*	stinging nettle	yes ($P < 0.001$)	no
speckled wood	*Pararge aegaria*	grasses	yes ($P < 0.001$)	yes ($P < 0.05$)
ringlet	*Aphantopus hyperantus*	grasses	yes ($P < 0.001$)	yes ($P < 0.05$)

The following two species in Table 2.8 have not increased their range but have increased in abundance (in fact, both species already occupy a very large range).

Table 2.8   Walk-and-count results from butterfly species that have not increased their range.

Common name	Scientific name	Larval food plant	Overall increase?	Increase in abundance in east higher than west?
green-veined white	*Pieris napi*	crucifers (cabbage family)	yes ($P < 0.001$)	yes ($P < 0.001$)
meadow brown	*Maniola jurtina*	grasses	yes ($P < 0.001$)	yes ($P < 0.001$)

Six other species in this category have also shown significant increases in the east compared to the west. These geographical trends are reinforced by the details of populations monitored for at least eight years (Figure 2.7). Note the tendency for increases and new colonizations in the east, although there are some exceptions.

○   What explanation can you offer for the overall increases in Tables 2.7 and 2.8?

●   It may be that the species have benefited from increases in their larval food plant. Nettles, grasses and members of the cabbage family are widespread and likely to have increased owing to increased roadside habitat, relaxation of intensive farming practices and increased suburban wildlife habitats (e.g. in gardens).

○   Why should there have been disproportionate increases in the east of the country?

●   This may be because there have been disproportionate increases in the factors listed above (e.g. relaxation of the previous higher levels of intensification in East Anglia) or perhaps owing to more pronounced regional climate differences.

The authors of this study attributed the increases in range to the larval food plant and suggested that reduced pollution events in the east may be important. As yet, there is no clear explanation for the geographical variation. The important point is that without this monitoring we could not begin to consider the underlying causes of the change.

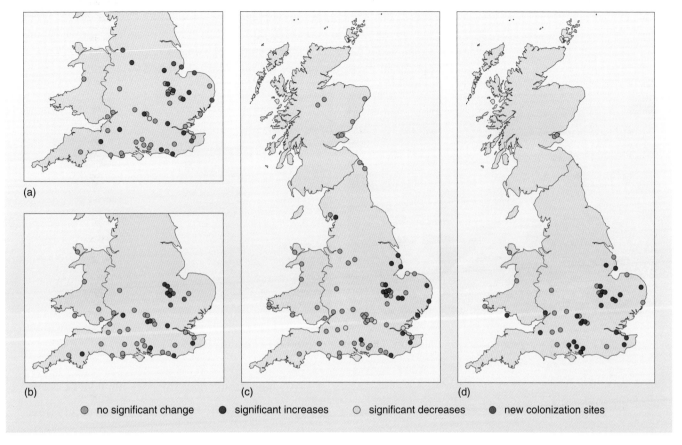

| ● no significant change | ● significant increases | ○ significant decreases | ● new colonization sites |

Figure 2.7 Example of changes found in walk-and-count sampling for four butterfly species: (a) comma; (b) speckled wood; (c) meadow brown and (d) ringlet.

The walk-and-count method has also been used in tropical regions.

○ What difficulties do you anticipate in using the walk-and-count method in tropical regions?

● The recorder will encounter a larger number of species that may have similar coloration. Individuals may fly at different times of day and may be difficult to see in forest areas. While the last three problems are not confined to tropical areas they are exacerbated in such habitats.

Despite these problems, it is still possible to collect useful data on tropical butterfly species. For example, Robin Bain and the author have been working on the responses of butterflies to local climate change up a mountain in Puerto Rico. The forest of El Yunque in eastern Puerto Rico (also known as the Luquillo Experimental Forest or Caribbean National Forest) has been the subject of climate study for more than 30 years. The three highest peaks reach approximately 1050 m. The tops of these mountains are covered with cloud forest (from an altitude of about 900 m) in which the dominant trees reach heights of only 2–3 m (Figure 2.8b). The cloud forest is threatened by lifting of the cloud base associated with deforestation in the lowland areas of Puerto Rico. Much of the island was logged in the 19th and early 20th centuries, and this process is now being repeated largely due to the pressures of tourism.

Puerto Rico has a number of endemic species, including 230 species of plants, 14 species of birds and a number of butterflies, including *Calisto nubila* (Figure 2.8a). The genus *Calisto* is restricted to the Caribbean islands. Using walk-and-count methods we have shown that, in the El Yunque region, *Calisto nubila* predominates in the upper part of the mountain (Figure 2.9).

(a)

(b)

Figure 2.8   (a) *Calisto nubila* and (b) cloud forest habitat at the top of El Yunque, Puerto Rico.

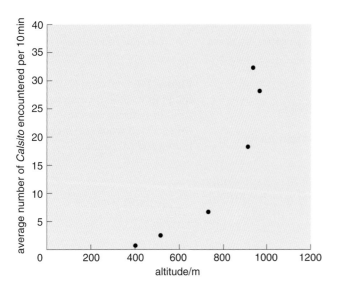

Figure 2.9   Change in abundance of *Calisto nubila* (a Puerto Rican endemic butterfly) with altitude.

○   At what altitude does the peak abundance occur?

●   At approximately 950 m, within the cloud forest habitat.

Predicting the effect of continuing shifts in the cloud base and alterations in climate, on *Calisto* and other species, will require more biological details, such as the thermoregulatory properties of the butterfly and its dependence on particular plants.

2.3.4   Sampling by indirect methods

An alternative to the direct counting of animals in a sample is the collection of evidence of their activities.

○   What kind of evidence of animal activity might provide information on the abundance of such animals?

●   Faecal pellets, bark scraping or leaf chewing, can all provide information on the presence of individuals and, through sampling, an indirect estimate of the abundance of the animals that made or caused them.

It is a perverse fact that the largest land animal in the world, the African elephant (*Loxodonta africana*), is one whose populations sometimes need to be sampled indirectly, by counting the dung of the animal (Figure 2.10), to produce estimates of population size. In forest areas, sampling often has to be undertaken by recording the amount of dung. This method has been applied to other endangered species, such as the giant panda in bamboo forests in China. In addition to using dung counts as estimates of population size, examination of the contents of dung (or regurgitated food such as owl pellets), may be used to determine the diet of animals.

Two important pieces of information are required when estimating elephant numbers based on the number of droppings. The first is the number of droppings produced per elephant per day, which will vary with the age of the elephant and between seasons. For example, the defaecation rate for bull elephants varies from 31 droppings per animal per day in the rainy season, to about 12 per day in the dry season. The second piece of information required is the number of days over which dung has accumulated. If elephants stay in one place for a long time they may defaecate repeatedly in that area. If all these droppings were counted then there would be an overestimate of the number of elephants present. A knowledge of the time over which dung has accumulated makes it possible to correct for this overestimate.

Figure 2.10 Elephant dung.

Knowing the defaecation rate per elephant per day (R), the number of days over which droppings have accumulated (A), (taking into account decomposition rates, which vary throughout the year), and the number of (estimated) droppings in the whole area (D), a formula can be derived to estimate the number of elephants (N), in the whole area:

$$N = D/(R \times A)$$

In a study in the Kasungu National Park, Malawi, D was estimated as 8 267 000; R was measured as 17 droppings per elephant per day, and A was 188 days.

○ Using these values, calculate the value of N, the estimated number of elephants in the park.

● 2587. Given that D, R and A may all vary considerably, an estimate of 2000–3000 is probably reasonable.

○ Why is it only possible to *estimate* the number of droppings produced by elephants?

● The number of droppings can only be estimated because it is simply not possible to count all the dung in, say, the area of a national park. The amount of dung in the larger area is therefore estimated by sampling in smaller areas.

2.3.5 Sampling using traps

The key to successfully estimating the size of animal populations by trapping, is to understand the ecology and behaviour of the individual animals. For example, with small mammals such as mice and shrews, knowledge of their feeding preferences and patterns of movement, allows traps to be positioned in appropriate places, to be baited with something they like eating and to contain dry grass ('bedding') to ensure their survival whilst in the trap. Simple records of captures will provide some information on the variety of species present and the relative sizes of the populations. However, the *best* estimates of population size come from capturing animals, marking them, releasing them, and then recapturing a proportion of them at a later time. These **mark–release–recapture** methods provide data that allow an estimate to be made of the size of a

Figure 2.11 An example of butterfly trapping, marking and release. (1) The butterfly is attracted to the fruit, feeds and then flies up into the net. (2) The butterfly is removed, held gently under a net and given a mark consisting of one or more dots, using felt-tip pen. (3) The dots represent the numbers 1, 2, 4, 7 and 10, 20, 40, 70. (4) Combinations of these numbers (dots), give the butterfly a unique number, e.g. the number 26 is represented by 2, 4 and 20. (5) The marked butterfly is then released, and possibly recaptured (6) (in the same or another trap).

population. If each animal is given a unique mark, then estimates can also be made of the distance covered by that individual (if it is caught in different traps), and its longevity.

The basic steps in any mark–release–recapture sampling are outlined here:

1 Animals are caught in traps, or by some other method, over a short period of time, e.g. a single night.

2 All these animals are marked in some way and then released. The number of caught and marked animals is recorded. It is important that neither the mark nor the trapping affect the behaviour or survival of the animal. Although this is an assumption of mark–release–recapture techniques, it is not always satisfied.

3 Traps are reset, and some animals are recaptured, perhaps on the night following marking and release. Two pieces of information are recorded: the total number of captures, and the number of recaptured animals, i.e. those marked on the first trapping occasion.

An example of a mark–release–recapture system for tropical butterflies attracted to fruit is given in Figure 2.11.

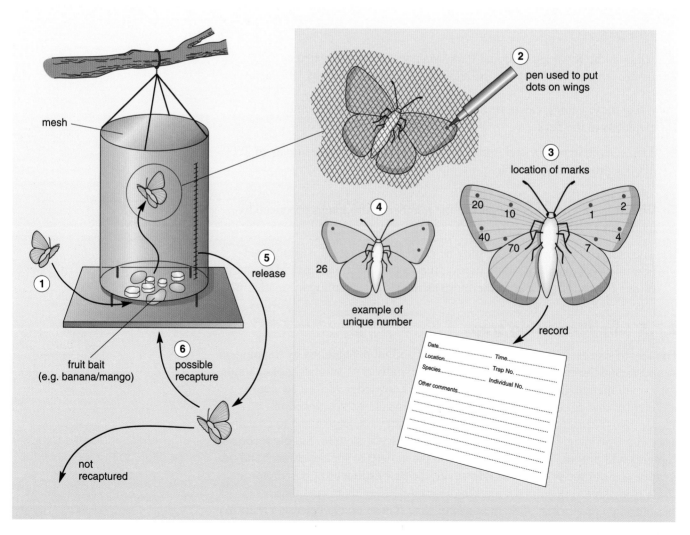

Knowing the fraction of recaptured animals allows an estimate to be made of the total population size. Suppose that 20 animals are caught on the first occasion (N_1, Figure 2.12). These are marked and released. On the second occasion 30 animals are caught (N_2) of which 10 are marked from the first sample (M_2). Therefore, $\frac{10}{30}$ or one-third of the second sample are recaptures. We know that 20 animals were originally marked. If they had mixed back evenly into the total population then we would expect that the fraction of marked individuals in the second sample (one-third) should be equal to the fraction of marked individuals in the total population. Therefore if the total population is T, then $20/T = \frac{1}{3}$. So $T = 3 \times 20 = 60$ individuals. This is the basis of many mark–release–recapture techniques and is summarized in Figure 2.12.

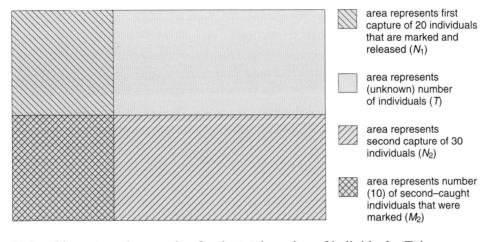

area represents first capture of 20 individuals that are marked and released (N_1)

area represents (unknown) number of individuals (T)

area represents second capture of 30 individuals (N_2)

area represents number (10) of second–caught individuals that were marked (M_2)

Figure 2.12 Concept of mark–release–recapture illustrated by overlap of rectangles. The total number (T) is unknown. 20 animals (N_1) are caught on the first occasion and marked. 30 animals are caught on the second occasion (N_2), of which 10 are found to have been marked (M_2).

Using this system, the equation for the total number of individuals *(T)* in a population is:

$$T = (N_1 \times N_2)/M_2$$

where the number of individuals caught and marked on a first occasion = N_1; the total number of individuals caught on a second occasion = N_2; and the number of marked individuals captured on a second occasion = M_2.

○ If 48 animals were originally marked, and 100 captured on a second occasion, of which 15 were marked, what is the total population size (T)?

● $T = (48 \times 100)/15 = 320$

Remember that this method can only provide an *estimate* of the size of populations; however, it has been used successfully on a wide variety of animal species, including insects, snails and small mammals.

2.4   Using data on population size

2.4.1 The likelihood of extinction

Once data on population size have been collected, how might they be used to identify populations or species that require protection or management? Plotting the *change* in population size over time (as we did for the green-winged orchid) allows us to see trends in the data. Statistical techniques can then be applied to quantify those trends. It is important to distinguish between three types of extinction event with reference to change in population size over time.

1   The first type of extinction, which is referred to below, depends only on fluctuations in the population size (it can be assumed that there is no change in the mean population size, Figure 2.13a), perhaps caused by external chance events such as weather patterns.

2   The second type of extinction occurs because the mean population size reduces over time, such that the population is driven towards extinction (Figure 2.13b). This may be caused by a variety of factors, such as habitat loss.

3   The final type of extinction combines the first two types, i.e. a population with a declining mean size, which also fluctuates about the mean (Figure 2.13c).

A useful variable when considering the types of extinction is the **probability of extinction** for a population. Probability quantifies the chance of an event occurring. A coin tossed in the air has a probability of $\frac{1}{2}$ of landing heads up, and $\frac{1}{2}$ of landing tails up. The probability of a rolled die landing with a 3 uppermost is $\frac{1}{6}$. We can define probabilities of extinction for populations in the same way. A population might have a 1 in 10 chance of becoming extinct in any one year, so its probability of extinction is $\frac{1}{10}$. A species with populations that have a high probability of extinction is an obvious candidate for conservation.

Why is it necessary to define a probability of extinction? The simple answer is that the natural fluctuations in population size over time mean that extinction is uncertain. In any year there is the possibility that the fluctuation will be sufficiently large to cause the population to become extinct. An estimate of the probability of extinction allows us to quantify that possibility. (Of course these fluctuations can result in the population size increasing as well as decreasing.)

With real populations it is very difficult to determine the probability of extinction.

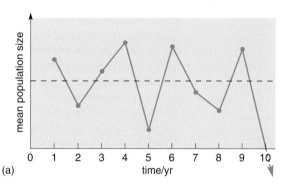
(a)

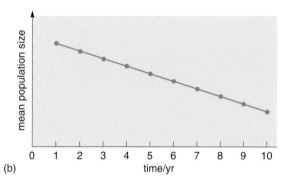
(b)

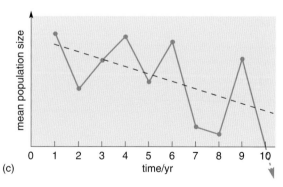
(c)

Figure 2.13   Three types of extinction event for populations over time. (See text for details.)

○   How can we calculate a probability of population extinction in the field?

● The only way is to monitor a number of populations over long periods of time. The proportion that become extinct in each time period would then be recorded and used to estimate the probability of extinction. This means that long-term detailed population studies are needed, repeated at different sites.

There are ways around this problem using mathematical modelling and statistical techniques. For example, the probability of an extinction event occurring in any one year can be related to the mean of the population size, and the variation around that mean (given by the variance or standard deviation), estimated from population sizes over time.

○ Calculate the mean population size (number of genets) and standard deviation of the green-winged orchid in Pilch Field, using the data in Table 2.6.

● The mean is 110.5 and the standard deviation is 31.1.

○ Imagine two populations with approximately equal mean size, but very different standard deviations. Which population would have the higher probability of extinction in any one year?

● The population with the higher standard deviation will have a higher probability of extinction, because its fluctuations are larger, leading to low population sizes from time to time, and therefore a greater risk of extinction.

○ Imagine two populations with approximately equal standard deviations, but very different mean values. Which population will have the higher probability of extinction?

● The population with the lower mean value, because the same fluctuations (equal standard deviation) will result in lower population values.

2.4.2 Population viability analysis (PVA)

The application of population dynamics (change in population size in time and/or space) to the conservation of rare species has come to be known as **population viability analysis (PVA)**. One of the aims of PVA is to estimate the probability of extinction. Much of the early work in the 1980s also stressed the need to identify the **minimum viable population size (MVP)** for a particular species. The definition of MVP appears both straightforward and appealing — what number of individuals of a particular species are required to enable a population to persist? Below this number the population will decline to extinction; above this number the population will persist.

It has frequently been said that MVP has no single value, or 'magic number', that has universal validity. This is, of course, correct. Indeed, the fact that we need to estimate a *probability* of extinction means that there *cannot* be one magic number! We could define MVP as the (initial) population size that reduces the probability of extinction (over a given period of time) to an 'acceptably' low level (e.g. a 1 in 100 chance of becoming extinct in 20 years). Defined in this way, MVP is determined by both the intrinsic dynamics of the species and the pattern of variation in climatic conditions or other extrinsic factors impacting on the population.

Despite the problem of calculating MVP, it is interesting to consider the sizes of populations recommended as minimally viable. For example, work on the grizzly bear in North America has suggested a MVP of about 50, based on considerations of fluctuations in fecundity and mortality. Other workers have suggested values of 500 or more for other species, dependent upon the amount of variation in population size. Many of these estimates have come from mathematical modelling work.

If a MVP value can be estimated, it could lead to a number of possible abuses. Any harvestable animal or plant species above its MVP may be regarded as available for exploitation as a biological resource. Similarly, habitat destruction could be justified *if* it was predicted that it still allowed a selection of species to remain above their MVP. Dilemmas arising from the concept of MVP and the probability of extinction abound. Conservation managers and organizations, presented with such numbers, have to make decisions. Should they divert their efforts to those species closest to the MVP, ignoring those that fall below it? Or should they let species near to or above the MVP get on with it and focus on the species below? Finally, it is worth remembering that, by its very nature, the probability of extinction is no guarantee for the practitioner (or the threatened population!). Hypothetically, if one were faced with the last herd of white rhinoceros (*Ceratotherium simum*) in Africa, would it be helpful to know that, based on the results of a mathematical model, the population had a one in four chance of becoming extinct?

The final part of this section will illustrate how studies of population size and probability of extinction have become incorporated into international conservation efforts.

2.4.3 Constructing practical measures of species vulnerability

It is clear that some assessment of species vulnerability is required which takes into account the size and dynamics of populations. Such an assessment has been developed by the International Union for Conservation of Nature and Natural Resources (IUCN). Species are placed in a **Red List of Threatened Species** within a series of categories, which includes the following five categories of extinction and threat (the IUCN categories refer to a general taxonomic unit, a taxon, rather than a species, but in most cases species is meant).

There are two categories of extinction:

Extinct   There is no reasonable doubt that the last individual has died (e.g. the passenger pigeon).

Extinct in the wild   Known to survive only in captivity, or as a naturalized population well outside its past range.

There are three threatened categories:

Critically endangered   Facing an extremely high risk of extinction in the wild in the immediate future.

Endangered   Not critically endangered, but facing a very high risk of extinction in the wild in the near future.

Vulnerable   Not critically endangered or endangered, but facing a high risk of extinction in the wild in the medium-term future.

You will see that the threatened categories are ranked in terms of the risk of extinction (represented by extremely high, very high or high over different time-scales of immediate, near or medium-term). An important step forward has been to use population data to begin to quantify these risks. The quantities include the levels of population reduction, the geographical range of the species and the population size. Species are put into categories on the basis of meeting one or more of a number of criteria based on these quantities. For example, a species is defined as critically endangered if it numbers less than 250 mature individuals, or has shown a reduction of over 80% within the last 10 years. By contrast, an endangered species would number less than 2500 mature individuals, or have shown a reduction of 50% over the last 10 years. Probability of extinction is also used. Thus a species for which quantitative analysis has shown a probability of extinction of at least 50% within 10 years, or three generations (which ever is the longer), is placed into the critically endangered category. The Red List also identifies species for which there are insufficient data to assign them to a category.

The Red List data are now so extensive that they are only provided in electronic format via the internet. Examples of species from the year 2000 list for the Caribbean islands are given in Table 2.9 and Figure 2.14. Some of the species are also found on the mainland. This list is based on studies of a number of taxonomic groups, e.g. parrots, flowering plants and mammals.

Table 2.9   Examples from the 2000 IUCN Red List of Threatened Species from the Caribbean islands (including, in some cases, species also found on the mainland).

Threat category	Species name	Common name or species description	Total number in category
extinct	*Amazona violacea*	Guadeloupe parrot	
	Ara tricolor	Cuban macaw	58
	Procyon gloveralleni	Barbados raccoon	
extinct in the wild	*Erythroxylum echinodendron*	tree	1
critically endangered	*Amazona vittata*	Puerto Rican parrot	142
	Calyptranthes kiaerskovii	tree	
endangered	*Guaiacum officinale*	lignum vitae (tree)	191
	Maytenus cymosa	tree	
vulnerable	*Crocodylus acutus*	American crocodile	
	Zanthoxylum flavum	tree	296

Figure 2.14   *Zanthoxylum flavum*, a Red List tree species in the Caribbean.

2.5   Summary of Section 2

1   A first step in the monitoring of species is the assessment of the geographical range of that species.

2   Global biodiversity hotspots focus on species with restricted geographical ranges (endemics). More than 20 such hotspots have been identified for a range of taxonomic groups. High percentage endemism may be caused by geographical isolation.

3   The steps in sampling a population as a means of assessing population size include choosing a site(s) (perhaps using distribution data), and then sampling directly or indirectly, often using traps for small mobile animals. Detailed examples are given for orchids, butterflies (including the Butterfly Monitoring Scheme) and elephants.

4   Population sizes (changing in time and/or space), can be summarized quantitatively using the mean and variance. The likelihood of extinction can be quantified by the probability of population extinction.

5   Minimum viable population size (MVP) can be defined with respect to the probability of extinction; for example, a minimum viable population is one that reduces the probability of extinction (over a given time) to an acceptably low level. This is part of the process of population viability analysis (PVA). These ideas have been incorporated into international conservation approaches to classifying endangered species, specifically the IUCN Red List of Threatened Species.

Management of habitats

3

The link between declining numbers of species and habitat loss or degradation has been emphasized in Section 1. When faced with a species whose numbers are decreasing (as detected by monitoring, Section 2), the first step is to identify the cause(s) of decline. Armed with that knowledge, this decline then needs to be reversed, or its rate reduced. For some species it may be simply enough to stop the current threat, such as deforestation, and allow the habitat, and therefore the particular species, to recover. For other species and habitats, a more active approach is required. Specifically, the species and/or its habitat need to be *managed* in order to restore population size to 'acceptable' levels (as discussed in Section 2). Most frequently it is management of the habitat, rather than management of the species, that is most appropriate. An advantage of managing the habitat is that one may help not only the targeted species, but also other species resident in that habitat.

3.1 Habitat management and secondary succession

Habitat management is a daily routine for many people involved in conservation in Britain and elsewhere. What many of them are attempting to do is to keep ecological succession in check. You will recall that succession is the directional change in the composition of plant and animal species over time, from habitats dominated by grasses and other small plants, to woodland dominated by large trees. In fact there are two types of succession that were covered in *Life* (Section 4.2).

○ What are the two basic types of succession?

● Primary and secondary succession.

○ Under what conditions do primary and secondary succession occur?

● A primary succession is one that begins on a substrate with no soil, an example being a newly formed volcanic island or a disused quarry. Secondary succession begins on soil cleared of its original vegetation. The methods of clearance include physical removal of vegetation (e.g. logging or ploughing) and fire (Figure 3.1), and influence the rate and course of succession.

Figure 3.1 Fire helping to maintain savanna grassland in southwest Guyana. Some of these fires are started by Amerindians to maintain the habitat at an early successional stage for cattle and goats.

This section covers two examples of habitat management and the relationship of the managed habitats to the natural state. The management required is frequently very simple and may appear crude, such as 'scrub-bashing', which is the removal of scrub to retain a grassland habitat. Similarly, use of grazing stock can keep potentially dominant grasses down and stop invasion by tree seedlings. Mowing the lawn is a method of management that stops coarse grasses and woody species invading and thereby prevents secondary succession.

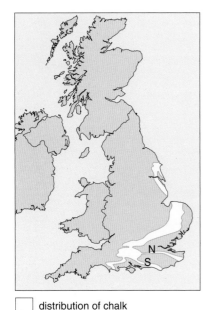

distribution of chalk

Figure 3.2 Distribution of chalk in Britain. (N, North Downs; S, South Downs)

3.2 Management of early successional habitats

3.2.1 Introduction

In this section we consider the management of various early successional habitats, principally grasslands. Consider, for example, the chalk grasslands of England, such as those on the North and South Downs (Figures 3.2 and 3.3), which are known to be amongst the richest habitats in Britain in terms of numbers of plant species per unit area. They are also important for a range of invertebrate species, such as the silver-spotted skipper (*Hesperia comma*). The natural vegetation of these areas was mature woodland (Section 1). In the absence of management, grassland on chalk will revert back to woodland because of secondary succession. This was seen in a 'natural experiment' that occurred in the 1950s. To understand this natural experiment and the particular form of management it reveals, it is necessary to consider the effect that rabbits have on vegetation on the Chalk and elsewhere.

Figure 3.3 A typical view of chalk grassland on the South Downs.

Figure 3.4 Rabbits are important grazers of downland.

3.2.2 Rabbits as agents of vegetation change

For many years, rabbits (Figure 3.4) have been considered agents of vegetation change. Their effects on vegetation are summarized in Extract 3.1, taken from the introduction to a scientific paper. Before you read the extract, note that Breckland is a large area of open grassland on sandy and chalky soils in East Anglia. *Suaeda fruticosa* is now known as *Suaeda vera* or shrubby seablite, *Calluna vulgaris* is a heather, and *Carex arenaria* is a sedge.

Extract 3.1
A. S. Thomas, Journal of Ecology (1960) 48, pp. 287–306.

Since 1900, there have been many studies of the influence of rabbit grazing on wild vegetation in Britain. One of the earliest accounts is by Wallis in *The Natural History of Cambridgeshire* (Marr and Shipley, 1904). Wallis described how the rabbits influenced the vegetation of Breckland, both by eating and by burrowing, and mentions some of the annual plants which grow on the soil disturbed by rabbits.

Oliver (1913) and Rowan (1913) described the feeding habits of rabbits on Blakeney Point, and listed plant species which were most attacked. In some cases the leaves were eaten, in others the rhizomes; and the branches of *Suaeda fruticosa* were bitten off.

Farrow (1913, 1917) carefully described and photographed the effects of rabbits on the vegetation of Breckland, showing how *Calluna vulgaris* was killed by grazing and replaced by heath of *Carex arenaria* and of grasses. Watt (1937) queried the importance attributed by Farrow to rabbits in determining the present aspect of the Breckland region.

Another notable study of the effects of rabbits on vegetation was that by Tansley and Adamson (1925) on the chalk of the South Downs, and in which were listed the plants eaten by rabbits, and the plants which were avoided by rabbits, and which therefore tended to increase when they were abundant. This study was extended by Hope-Simpson (1940).

Other notable studies of the effects of rabbits were those of Fenton (1940) on hill grazing in Scotland; of Phillips (1935) on a reseeded pasture in Wales; and Gillham (1955) on the island of Skokholm, where vegetation under an abundance of rabbits was contrasted with that of the ungrazed island of Grassholm.

A prolonged and detailed study of the effect of excluding rabbits from Breckland grassland by Watt (1957) recognizes the possible effects of using wire-netting enclosures, and shows that the change from grazed to ungrazed is not likely to follow a straight course when gains or losses are plotted against time.

Extract 3.1 introduces two important effects of rabbits that impact on a variety of early successional habitats:

1   The plants of some species are preferentially eaten, reducing their abundance in rabbit-grazed vegetation. Conversely, some plants are avoided by rabbits and so tend to increase when rabbits are abundant. The result may be a shift in community composition, such as that reported by Farrow, from a heather-dominated community to a grass/sedge-dominated community.

2   Rabbits create gaps in the vegetation ('soil disturbed by rabbits') within which annual plants may grow.

○   As annual plants are characteristic of the early stages of secondary succession, what is the effect of rabbits on secondary succession?

●   Rabbits, in creating conditions suitable for early successional plants, are slowing down or even stopping succession.

A third factor, not mentioned in Extract 3.1, is that rabbits, at suitable densities, will tend to reduce the height and quantity of the vegetation to a low level. This will have an adverse effect on some later successional plants, such as coarse perennial grasses and tree seedlings, which, because they are better competitors, eventually come to dominate the vegetation in the absence of grazing. Tree seedlings, which as adults will dominate later successional stages, and whose growing points are not close to the ground, may be killed by grazing rabbits. This emphasizes the effect of rabbits in secondary succession, not only slowing it down but possibly keeping it at a *false* climax (end-point), referred to as a plagioclimax or deflected climax (*Life* Section 4.2.5).

Deflected climax communities on chalk grassland affected by rabbits are typified by very short-clipped vegetation, scrape holes and few trees. Because no single plant species is able to dominate, the number of plant species per unit area may be very high, perhaps 30–40 species in $1 \, m^2$. We might therefore consider rabbits, or other grazers, as potential management tools for maintaining chalk grassland, creating conditions suitable for high numbers of plant species and favouring early successional plant species which otherwise might be quite rare. Convincing evidence of the effect of rabbits came in the 1950s when the rabbit population in Britain suffered a major decline, as revealed in Extract 3.2, which you should read now.

Extract 3.2
A. Tittensor, Country Life, 3 February 1983.

Myxomatosis became familiar to most of us three decades ago when it swept rapidly through our rabbit populations, reducing them to a mere remnant of their former abundance. It is still with us, but both rabbit and disease have adapted to meet changing circumstances, and a temporary balance has been reached.

Myxomatosis is caused by the myxoma virus, and first caught the attention of scientists in 1896 when a stock of domestic laboratory rabbits in Montevideo was almost eliminated by an outbreak of the disease.

The introduction of myxomatosis to Europe and Australasia in order to reduce rabbit numbers was suggested as early as 1918, and field trials were started in the mid-1930s. There was, for example, an unsuccessful series of experiments on the Welsh island of Skokholm between 1936 and 1938. These early trials failed because the importance of blood-sucking insects, particularly fleas and mosquitoes, in transmitting the disease was not appreciated until later. It was subsequently discovered that virus particles are transferred on the insects' mouthparts from infected rabbit to new host. We now know that Skokholm was an unfortunate choice for such experiments, since it is the only known place in Britain where rabbits have no fleas.

Myxomatosis reached Europe in June 1952, when a retired doctor released two rabbits inoculated with virus on his sporting estate near Chartres in France. The estate was walled in, so it was anticipated that the disease would be confined to the site of introduction. However, it reached the Channel coast in December 1952, extended to southern France by June 1953, and had reached much of western Europe by the end of 1953. Myxomatosis crossed the Channel to an estate near Edenbridge in Kent during 1953, and had spread through much of south-eastern England by the following February, despite government efforts to eradicate it at source. Within two years the disease had reached most parts of Britain.

The resultant effect on the vegetation, following the crash in rabbit numbers, was dramatic, as revealed in the 'Discussion' section of A. S. Thomas' 1960 paper, part of which is reproduced here as Extract 3.3. In Extract 3.3, note that Kingley Vale, Old Winchester Hill and Pewsey Vale all contain chalk grassland habitats, and *Juniperus* and *Taxus* are the genera of juniper and yew respectively. *Crataegus monogyna* is hawthorn, *Atropa belladonna* is deadly nightshade, *Cynoglossum officinale* is hound's-tongue, and *Senecio jacobaea* is ragwort. Some of these plant species are illustrated in Figure 3.5.

(a)

Extract 3.3
A. S. Thomas (1960) *Journal of Ecology*, 48, pp. 287–306.

Spectacular changes in vegetation since the death of the rabbit have been reported from other countries: Australia (Ratcliffe, 1956), France (Morel, 1956), and the Netherlands (van Leeuwen, 1956). In the main, the changes reported in this paper were similar to those found in other countries, especially in the better regeneration of woody plants.

(b)

The bad influence of the rabbits in damaging and killing planted trees has long been known to foresters, and great expense has been incurred to erect rabbit-proof fences. But the localized influence of rabbits only becomes obvious in the course of this work, for it has determined the present aspect of some Nature Reserves. For example, Watt (1926) carefully mapped Kingley Vale, showing areas of *Juniperus* and *Taxus* scrub; he noticed that the rabbits grazed both *Juniperus* and *Taxus* and thought that their selective action benefited *Juniperus*. But these areas are now almost pure *Taxus* scrub, scarcely a *Juniperus* plant remaining. It has been noticed that large bushes of *Juniperus* tended to die back in the presence of rabbits, even though their bark was not eaten and their young shoots were above reach of the animals. As mentioned above, after the rabbits died many young *Juniperus* bushes became evident in the grassland, and it may be that this plant will become common again in the south of England.

(c)

Crataegus monogyna also was greatly curtailed by rabbits. For example, around the warren near the northern corner of the Old Winchester Hill Nature Reserve, there were few small *Crataegus* bushes and many *Taxus* bushes; but on the south-west side of the reserve, where rabbits were much fewer, dense *Crataegus* scrub had grown up. Therefore it is likely that, where the grazing influence of rabbits is not replaced by the grazing influence of sheep and cattle, scrub of *Crataegus* and other shrubs and trees may rapidly colonize much of the downlands.

(d)

Observers in other countries have emphasized the increase in various plants since the death of the rabbits, but few have reported such a decrease as shown by field notes on some of the transects in the south of England. It is only to be expected that plants encouraged by rabbits have become fewer; that will be a boon to farmers because many of these plants, such as *Atropa belladonna* and *Cynoglossum officinale,* are poisonous to stock. *Senecio jacobaea* undoubtedly was encouraged by rabbits, who exposed and disturbed the soil by grazing and scratching, enabling the plant to become established. Although rabbits were generally considered to avoid *S. jacobaea* (Harper and Wood, 1957) yet it was observed in several places still infested by rabbits in 1954 that they prevented some *S. jacobaea* plants from fruiting, for they gnawed and felled the flowering stems, but ate little of them. After the rabbits died in 1954 and 1955 there was a spectacular display of flowering *S. jacobaea*. Colour photographs of the Pewsey Vale escarpment in Wiltshire showed patches of *S. jacobaea* only on rabbit burys (areas of soil disturbed by rabbits) in 1954, but a sheet of yellow flowers over the hillsides in 1955; in 1956 scarcely a single plant of *S. jacobaea* flowered there; in 1957 there were few plants in flower, again on the burys.

(e)

Figure 3.5   Plant species which are affected by rabbit grazing in Extract 3.3: (a) juniper; (b) yew; (c) hawthorn; (d) deadly nightshade and (e) ragwort.

The decline in rabbit numbers due to myxomatosis revealed how the rabbit population had been affecting the composition and structure of the natural vegetation. By reducing the rabbits to a low density, a 'natural experiment' had occurred that demonstrated the effect of these grazers on vegetation. It also revealed the rabbits' efficacy, and potentially that of other grazers, as management tools for chalk grassland.

An interesting dilemma for conservation is posed by this result. Juniper, a native and uncommon plant in need of conservation, seemed to thrive after rabbit grazing declined. So, if management such as grazing, which can increase plant species richness, is implemented, it may be done at the expense of species such as juniper. In other words, the increase in numbers of plant species is *selective* in favour of certain early successional species. Similarly, if large numbers of plants in flower are desired, intensive uninterrupted grazing may not give the required result. These dilemmas abound whenever management options are considered.

3.2.3 Measurement of the diversity of communities

An important feature of chalk grassland is the high number of species in a given area. It is tempting to describe it as a high diversity system, providing a clear mandate for management to maintain or even increase the diversity. However, the term diversity has a particular meaning in ecology. Once that meaning is understood, diversity measures can provide a useful tool with which to monitor the community and assess the efficacy of management.

○   Recall from *Life*, the meaning of species diversity.

●   Diversity is a measure of the richness and evenness in composition of the units of an ecological system (*Life* Section 4.3.1).

Given that an ecological community is defined as being composed of individuals of different species (*Life* Section 2.1); a community with *high* biological diversity is one that has about equal numbers of individuals (high evenness) of many different species (high species richness). It is not the total number of individuals that is important in determining species diversity, but rather the relative proportions of individuals of each species and the number of species. There are several different types of diversity index, each of which aims to summarize the diversity of a community (or other ecological system) in a single value. In the following example, one of the simpler measures of diversity, the Simpson index, will be used.

The Simpson diversity index

An attractive property of the Simpson diversity index, is that the maximum theoretical value (highest diversity) is very close to one (although it can never equal one) and the minimum theoretical value is zero. To calculate the Simpson diversity index for any community we need to know the number (or biomass) of individuals of each species in that community. We then calculate the fraction of the total number of individuals contributed by each species. For example, if there are 30 individuals in an area and 10 of these are species A, then the fraction of 'A' individuals is $\frac{10}{30}$ or $\frac{1}{3}$. The fraction for each species is then squared, the squared values are totalled and the total subtracted from 1 to give the Simpson index value. The following examples should make these calculations clearer and illustrate what determines whether a community has a high or low diversity.

Let us begin by considering a hypothetical community made up of four species, each with ten individuals (Table 3.1).

Table 3.1 Hypothetical community of 40 individuals of four equally abundant species.

Species	Number of individuals	Fraction of total	(Fraction)^2
A	10	$\frac{1}{4}$	$\frac{1}{16}$
B	10	$\frac{1}{4}$	$\frac{1}{16}$
C	10	$\frac{1}{4}$	$\frac{1}{16}$
D	10	$\frac{1}{4}$	$\frac{1}{16}$
total	40	1	$\frac{1}{4}$

Simpson diversity index value is $1 - \text{total (fraction)}^2 = \frac{3}{4} = 0.75$.

So this community of four species has a Simpson diversity index of 0.75. Now compare this result with that obtained by keeping the same total number of individuals in equal proportions but doubling the number of species (Table 3.2).

Table 3.2 Hypothetical community of 40 individuals of eight equally abundant species.

Species	Number of individuals	Fraction of total	(Fraction)^2
A	5	$\frac{1}{8}$	$\frac{1}{64}$
B	5	$\frac{1}{8}$	$\frac{1}{64}$
C	5	$\frac{1}{8}$	$\frac{1}{64}$
D	5	$\frac{1}{8}$	$\frac{1}{64}$
E	5	$\frac{1}{8}$	$\frac{1}{64}$
F	5	$\frac{1}{8}$	$\frac{1}{64}$
G	5	$\frac{1}{8}$	$\frac{1}{64}$
H	5	$\frac{1}{8}$	$\frac{1}{64}$
total	40	1	$\frac{8}{64} = \frac{1}{8} = 0.125$

Simpson diversity index value is $1 - 0.125 = 0.875$.

The Simpson diversity index value has increased from 0.75 to 0.875. Now let us return to a four-species community with 40 individuals, except that this time the distribution of individuals is different between species (Table 3.3). In this case, species A is *dominant,* comprising 70% of the total.

Table 3.3 Hypothetical community of 40 individuals comprising four species, one of which is highly abundant.

Species	Number of individuals	Fraction of total	(Fraction)^2
A	28	$\frac{28}{40} = 0.7$	0.49
B	4	$\frac{4}{40} = 0.1$	0.01
C	4	0.1	0.01
D	4	0.1	0.01
total	40	1	0.52

Simpson diversity index value = $1 - 0.52 = 0.48$.

○ What has been the effect of altering the proportions away from equal values for the community of four species?

● It has *reduced* the Simpson diversity index from 0.75 (Table 3.1) to 0.48.

○ What is the relationship between dominance and diversity revealed in this example?

● Dominance is *inversely* related to diversity.

These examples show that high Simpson diversity index values are produced by higher numbers of species and a more even spread of individuals between species. Of course, the ability to quantify diversity is quite different from understanding *why* a particular community is highly diverse. This is a problem with which ecologists have grappled for many years and its discussion is largely beyond the scope of this topic, although we have noted the importance of factors such as grazing in maintaining a high diversity of plant species in chalk grassland.

3.3 The relationship between managed and natural woodland

In this section we switch our attention from early successional habitats to late successional habitats, i.e. woodland. Coppicing is a form of woodland management in which trees such as hazel (*Corylus avellana*) are cut down and allowed to re-sprout. The stems are harvested at certain time intervals (between 5 and 15 years) depending on the use of the wood. This practice results in newly open areas and areas of gradually closing canopy, i.e. a mosaic of conditions, which may favour a range of species such as the pearl-bordered fritillary (*Boloria euphrosyne*) and the high brown fritillary (*Argynmis adippe*, *Life* Figure 2.60). Coppicing illustrates the principle of maintaining the habitat at different successional stages. This section will focus less on the details of woodland management, and more on how managed woodlands compare to the natural state.

3.3.1 The natural state

In Section 1 we noted that the wildwood in Britain had been lost, except for possibly a few sites in Scotland. Indeed, there are very few examples of virgin forest (wildwood) remaining in Europe. Most of these are believed to have survived due to their inaccessibility. Thus the challenge is to find *any* woodland that may indicate the natural climax state towards which managed forests may be directed.

There are a number of criteria for determining virgin forest, in addition to consideration of vegetation structure and species composition. These were listed by the woodland ecologist, George Peterken and include:

- No sign of direct human interference.
- Full range of large animals, including carnivores, which are not significantly affected by events outside the site.
- Historical records should go back to the origins of settlement in the neighbourhood and show past interference is unlikely.
- Surrounding events should not significantly affect the site.

Parts of the Bialowieza Forest in Poland seem to meet some of these criteria, although the area has been affected by human activity over the last 200 years. Bialowieza Forest, meaning the forest of the white tower, covers 125 000 ha of a level lowland plain. It is the largest virgin forest in lowland Europe, surviving in a near natural state until the end of the 18th century. Since then it has been reduced by clearance and disturbed by felling and overgrazing. Within this area lies the Bialowieza National Park of 4747 ha, protected since 1921. It supports a good variety of large mammals, including the only wild herd of European bison (*Bison bonasus*, Figure 3.6). The original population numbered over 700 in the early 20th century but was destroyed by 1919. It was then re-established from animals surviving in zoos. The only other wild population of European bison at the beginning of the 20th century, in the Russian Caucasus, became extinct in 1927. Bialowieza also has populations of wolf, wild boar (Figure 3.6) and red deer, but some of the larger animals have been lost, including brown bear. Although an imperfect fit to Peterken's criteria, Bialowieza is probably the closest area to virgin forest in Europe (certainly in the lowlands). Accordingly, we shall spend a little while describing this site before making various comparisons.

(a)

(b)

Figure 3.6 (a) European bison (*Bison bonasus*) and (b) wild boar (*Sus scrofa*).

The park contains four types of woodland vegetation divided into deciduous and coniferous forest. The deciduous forest is composed of a community of pedunculate oak, small-leaved lime and hornbeam on higher ground, with mixtures of alder and ash, or ash and wych-elm along the streams. The coniferous forest is found in wet and acidic areas, including spruce woodland and pinewoods. It seems that the spruce woodland is being replaced by the deciduous woodland.

○ What evidence might there be for the replacement of coniferous woodland by deciduous woodland?

● The presence of seedlings and young trees of deciduous woodland within the areas of coniferous woodland (indeed, this is the case).

The biological diversity of Bialowieza is immense. There are 990 species of vascular plant, 254 species of bryophyte (mosses and liverworts) and 200 species of lichen. Of the 990 species of plant, 750 species are believed to be natural to the forest (i.e. not influenced by the activities of humans).

3.3.2 Comparisons with managed woodlands

Managed woodlands differ in a number of ways from the few remaining virgin forests (sometimes called old-growth forests). Generally they are smaller, with an increased amount of edge habitat. The type of disturbance within the woodland will also be different. Trees in an old-growth forest will fall naturally from old age, disease or wind-throw, creating gaps in the forest and leaving wood on the ground. This wood is then attacked by a variety of fungi and invertebrates. In managed woodland, disturbance may be created by coppicing for example, and wood may be removed. Trees may also have been planted. The history of the sites will be different. Managed woodlands may have been partially or wholly cleared at sometime in the past, or, at least, had a history of management. In contrast, old-growth forest should have a history of natural processes, although even in the case of Bialowieza there has been some interference in the past. We will now consider the effect of some of these differences on the woodland flora and fauna.

Many studies have shown the importance of area and edge effects on community composition. It is not a surprise to find that a large woodland like Bialowieza has a much greater number of species than a smaller woodland. In Table 3.4 the contrast is made with a well-studied woodland in Britain — Monks Wood National Nature Reserve in Cambridgeshire. Bialowieza has almost three times as many vascular plants or bryophytes (contrast columns 2 and 3) and six times as many lichens.

Table 3.4   Comparison of species richness of selected taxonomic groups between Monks Wood National Nature Reserve in southern England and Bialowieza Forest in Poland.

Type of plant	Monks Wood NNR (157 ha)	Whole Bialowieza Forest (125 000 ha)	150 ha sample of Bialowieza Forest
vascular plants	372	990	305
bryophytes	97	254	146
lichens	34	200	152
fungi	337	not known	1327

○   What major problem do you envisage in making the comparison between Bialowieza Forest and Monks Wood?

●   The main problem is that Monks Wood is in Britain, which has a smaller total number of species than mainland Europe.

A more interesting comparison, which does not avoid this problem of geographical variation, is the comparison between a sample of Bialowieza Forest of the same size as Monks Wood (Table 3.4, columns 4 and 2).

○ How do the four groups of species vary in the differences between the two sites?

● The number of vascular plants is actually lower in the sample of Bialowieza Forest. This is in contrast to the other three taxonomic groups, which are higher in the sample of Bialowieza Forest. The increase in number of lichens and fungi is striking.

It may be that the increased number of vascular plants in Monks Wood compared to the sample is due to edge effects, i.e. that a substantial proportion of the 372 species are not true woodland species but woodland edge and grassland plants. Let us develop these ideas with reference to the bird fauna in Bialowieza Forest.

Bialowieza Forest contains 226 species of bird (of which 169 breed in the area). Of these, 107 species are forest or forest edge species. In contrast, the whole of Britain only has 75 woodland associated species. The commonest bird species in Bialowieza Forest were the same in all forest types, including birds common across much of lowland Europe (chaffinch, robin, chiffchaff, wood warbler, blackcap, collared flycatcher, songthrush, hawfinch and great tit). Studies of birds in Bialowieza Forest reveal the effect of edge habitats (Table 3.5).

Table 3.5 Density and species richness of birds in Bialowieza Forest.

Forest type	Pairs per 10 ha (mean and range)	Number of species per plot* (range)
Interior		
riverine ash–elder	75 (65–87)	35–38
oak–lime–hornbeam	62 (55–72)	28–36
mixed conifer and broadleaved	34 (33–37)	28
pine–bilberry	36 (27–41)	25–27
Edge		
riverine ash–elder	100 (93–105)	51
oak–lime–hornbeam	78 (65–80)	42

* The size of plots was 24–32 ha.

○ How does bird density (number of pairs per 10 ha) and species richness alter between interior and edge habitats?

● Comparing within the same vegetation types shows an increase in density from interior to edge (oak–lime–hornbeam 62 to 78, riverine ash–elder 75 to 100). Species richness also increases from interior to edge, with the ash–elder forest gaining approximately 14 species and the oak–lime–hornbeam gaining a minimum of six species.

The increase of density and species richness in edge habitats is found in a wide variety of woodland examples. Even in tropical forest there are increases in disturbed or edge forest due to the influx of generalist species which can exist in a range of habitats. Another explanation of the increase in density in edge habitats is the decline in populations of predators. In the interior of Bialowieza Forest the density of birds was low owing to large populations of predators (ten species of eagle and hawk, five species of owl).

○ What are the main differences between the interior deciduous and coniferous woodland in Table 3.5?

● Coniferous woodland has a much lower density of species (approximately half that of the deciduous woodland). The difference in species richness is less apparent, although the coniferous species richness is at the lower end of the range of deciduous woodland values.

Understanding the species composition of woodlands depends in part on a historical perspective, e.g. the distribution and abundance of present-day plants may have been affected by past herbivore activity along with current herbivory. To illustrate this point, consider the list of feeding preferences of the large herbivorous mammals in Bialowieza (Table 3.6).

Table 3.6 Preferred and avoided plant species of three large herbivores in Bialowieza Forest.

	Bison	Boar	Deer
preferred plant species	lesser celandine (*Ranunculus ficaria*)	marsh marigold (*Caltha palustris*)	marsh marigold (*Caltha palustris*)
	hedge woundwort (*Stachys sylvatica*)	herb robert (*Geranium robertanium*)	meadowsweet (*Filipendula ulmaria*)
		stinging nettle (*Urtica dioica*)	creeping buttercup (*Ranunculus repens*)
avoided plant species	wood anemone (*Anemone nemorosa*)	wild garlic (*Allium ursinum*)	
	wood sorrel (*Oxalis acetosella*)		
	greater stitchwort (*Stellaria holostea*)		

These preferences may reflect the broad habitat preferences of the animals. For example, *Caltha palustris* (marsh marigold) and *Ranunculus repens* (creeping buttercup) are found in wetter areas. It is interesting that wild boar do not like wild garlic (*Allium ursinum*)! All of these plant species are found in woodlands and other habitats in Britain where at least two of these herbivores are not present. Thus Bialowieza Forest is not an entirely helpful model for biological conservation in Britain!

The simple message for conservation is that using virgin forest (or other near 'natural' ecological systems) is not a straightforward way to set objectives for management. Key 'natural' components, such as large herbivores, may be missing (and may have been missing for many hundreds of years). Similarly, it is likely that near natural areas are very large and that management of small areas requires different techniques. Fortunately, modern ecology provides a wealth of data and techniques to allow for different sizes of habitats. As stated earlier in this topic, humans are in the privileged position of choosing the fate of many of the species on the planet. Whilst large wilderness areas might be a poor model for much of modern conservation management, they at least represent a first approximation to what might be expected to be found in the area of a given habitat. And, of course, they are exceptionally worthy candidates for biological conservation in their own right.

3.4   Summary of Section 3

1   Habitat management often involves the control of secondary succession, for example slowing down or stopping succession at a false endpoint.

2   The study of the rabbit in Britain during the 20th century has revealed the potential for chalk grassland management by grazing animals.

3   Diversity of ecological communities can be measured by indices which combine species richness and relative abundance.

4   Managed woodland needs to be compared with woodland which is relatively undisturbed so that appropriate management objectives can be detailed. Bialowieza forest is described as an example of an extensive woodland in a near virgin state.

Learning outcomes for Topic 11

After working through this topic you should be able to:

1   Define the main conservation problems facing species and give examples of these.

2   Understand the need for sampling and monitoring of populations and give examples of both.

3   Appreciate the importance of size and fragmentation of populations and habitats in biological conservation.

4   Describe the relationship of management to secondary succession.

5   Understand the principles of diversity indices and make calculations of one of these.

6   Contrast managed habitats with habitats in near natural condition.

Acknowledgements for Topic 11 *Biological Conservation*

Grateful acknowledgement is made to the following sources for permission to reproduce material in this book:

Figure 1.2a: Stephen J. Krasemann/Science Photo Library; *Figure 1.2b*: Wayne Lawler/Ecoscene; *Figures 1.4, 1.8, 1.11, 1.14, 2.8, 2.11, 2.14, 3.1, 3.3, 3.5c*: Mike Gillman/Open University; *Figure 1.9a*: Copyright © 1993 Smithsonian Institute; *Figure 1.9b*: Mitchell Library, State Library of New South Wales; *Figures 1.9c and d, 2.2, 2.4, 2.5, 3.4, 3.5, 3.6*: Mike Dodd/Open University; *Figure 2.1a*: Bud Lehnhausen/Science Photo Library; *Figure 2.1b*: Pat and Tom Leeson/Science Photo Library; *Figure 2.1c*: TH Foto-Werbung/Science Photo Library; *Figure 2.10*: Copyright © 1998 Worldwander.com.

Every effort has been made to trace all the copyright owners, but if any has been inadvertently overlooked, the publishers will be pleased to make the necessary arrangements at the first opportunity.

Index

Note: Entries in **bold** are key terms. Page numbers referring to information that is given only in a figure or caption are printed in *italics*.

A

Acacia nilotica 16
adonis blue butterfly (*Lysandra bellargus*) 56
Afghanistan, droughts 11
Africa, grasslands of 15–16
African elephant (*Loxodonta africana*), estimation of numbers 146–7
Agouti paca (paca) 93, *94*
agricultural grasslands 13
agriculture, effects on grassland biodiversity 44–7
agrochemicals 45
agroecosystems 5–6
Agrostis spp. 29
Ailuropoda melanoleuca (giant panda) 135, 146
Alauda arvensis (skylark) 45, 46
alder (*Alnus glutinosa*) *30*, 31
Alouatta seniculus (red howler monkey) 95, *96*
alpha diversity 78, **79**
Amazon forests 73, 76, 78, 80, 82, 88, 91, 95
amenity grasslands 21
American bison (*Bison bison*) 14, 43
amphibians, global hotspots for *139*
ancient woodland *see* wildwood
Andropogon spp. 14, 15
Anemone nemorosa (wood anemone) 126
angwantibo (*Arctocebus calabarensis*) 88
antbirds (Formicariidae) 101, *102*
antbutterflies (Ithominae) 101–2
Anthriscus sylvestris (cow parsley) 32
Antilocapra americana (pronghorn) 14
ants 80, 81
    see also army ants; red ant
Arctocebus calabarensis (angwantibo) 88
Argynnis adippe (high brown fritillary) 38, 162
Aristida spp. 15
armadillo 15
army ants 101, *102*
Arrhenatherum elatius (false oat grass) 31, 32
Asian elephant 122
Asphodius niger (Beaulieu dung beetle) 41
Ateles paniscus (black spider monkey) 95, *96*
Atlantic Forest, Brazil 139, 140
Atropa belladonna (deadly nightshade) 159
Australia, grasslands of 16
autotrophs 98
avoidance mosaics 33, 37

B

Balaenoptera musculus (blue whale) 135
Barteria fistulosa 80
bastard toadflax (*Thesium humifusum*) 54
Batesian mimicry 102
bear *see* brown bear
Beaulieu dung beetle (*Asphodius niger*) 41
Bellis perennis (daisy) 33
Bertholletia excelsa (Brazil nut tree), farming 103
beta diversity 79
Betula spp., changes in abundance *131*, 132
Bialowieza Forest, Poland 163–6
biodiversity 6, 7
    effects of hunting on 43–4
    of tropical forests 77–84
    see also genetic diversity
biodiversity hotspots 132, 137–40
biological conservation *see* conservation
Biological Dynamics of Forest Fragment Project (BDFFP) 107
biomes 123–4
birches *see Betula* spp.
birds
    of calcareous grasslands 54–5
    global hotspots for *138*, 139
    of natural and managed woodlands 165–6
    population changes of British species 44, 46, 47
    of tropical forests 82, 85, 95
    of wet grasslands 58, 59–60, 62
Bison bison (American bison) 14, 43
Bison bonasus (European bison) 163, *166*
black spider monkey (*Ateles paniscus*) 95, *96*
blue whale (*Balaenoptera musculus*) 135
Boloria euphrosyne (pearl-bordered fritillary) 162
Bouteloua gracilis 14
brambles (*Rubus* spp.) 33
Brazil nut tree (*Bertholletia excelsa*), farming 103
Breckland 156, 157
breeding sites, partitioning in tropical forests 88–9
British grasslands 18–33
    composition of communities of 27–33
    conservation of 50–51
    history of 18–20
    types of 20–26

C

calcareous grassland 21, 23–4, 53–7, 128, 156, 158, 160
calcicolous plants 23–4, 127
calcifugous grassland 21, 24–6, 29
    species of 127
Caligo sp. (owl butterfly) 93, *94*, 102
Calisto nubile 146
Calluna vulgaris (heather) 156, 157
camouflage 102
campos 14–15
cane toad (*Bufo marina*) 44
Canis lupus (grey wolf) 14, 129–30
canopy of tropical forest 94–5, *96*
Carex arenaria (sedge) 156, 157
carnivores 98
cattle 55–6
cauliflory 82
Cebus capucinus (white-faced capuchin monkey) 110, *111*
Ceratotherium simum (white rhinoceros) 16
cereals 5
chalk grassland *see* calcareous grassland
chalk milkwort (*Polygala calcarea*) 54
Charaxinae 95, *96*
Chiltern gentian (*Gentianella germanica*) 54
climate
    effect on grassland distribution 10–12
    and habitat type 123–4
    and species richness 90
climatic climax vegetation 12
climax forest **71**
cloud forests 79, 145–6

Briza media (quaking grass) 28
bromeliads 81, 85, *87*
brown bear (*Ursus arctos*) 129–30
bryophytes in tropical forests 78
Bufo marina (cane toad) 44
bushbabies 88
butterflies
    cryptic coloration 93, *94*
    mark–release–recapture sampling of 148–9
    of tropical forests 93, *94*, 95, *96*, 107
    see also antbutterflies *and individual butterfly species*
Butterfly Monitoring Scheme 143–5
buttresses 81

coastal grasslands *see* wet grasslands
cocksfoot (*Dactylis glomerata*) 32
common reed (*Phragmites australis*) 30
communities
    of British grasslands 27–33
    measurement of diversity of 160–62
competitive exclusion principle 85
coniferous woodland *126*, 127, 163, 166
conifers in tropical forests 78
conservation 120, *121*
    of grasslands 49–53
    terminology 122
    of tropical forests 109–11
coppicing 162
coral snake (*Micrurius dissoleucus*) *94*
corridors 108
Corylus avellana (hazel) 162
Countryside Stewardship Scheme 53
cow parsley (*Anthriscus sylvestris*) 32
cowslip (*Primula veris*) 57
Crataegus monogyna (hawthorn) 159
cross-leaved heath (*Erica tetralix*) 29, 127
crypsis 93, *94*
Cuba, savanna in 13
Cynoglossum officinale (hound's tongue) 159
Cypripedium sp. (lady's slipper orchid) 135

D

Dactylis glomerata (cocksfoot) 32
daisy (*Bellis perennis*) 33
Damodar Valley, India, reforestation 108
dandelion (*Taraxacum* spp.) 32
Darwin, Charles 83
deadly nightshade (*Atropa belladonna*) 159
deflected climax (plagioclimax) vegetation
    12, 158
deforestation 12, **75**
    of Britain 124–8
    measuring 75–6
Dendrobates truncatus (poisonous arrow tree
    frog) *94*
detritivores, role in grassland ecosystems 41
detritus ecosystems 35, 99
dipterocarps in tropical forests 78
distribution gradients (grasses) 28
disturbed forests *see* secondary forests
diurnal species **85**
docks (*Rumex* spp.) 32
dodo 128
dog violet (*Viola riviniana*) 38
drip-tips 82
Duke of Burgundy butterfly (*Hamearis
    lucina*) 57
Dumboya sp. (wild pear) *8*

dung beetles 41
dung counts 146–7

E

early gentian (*Gentianella anglica*) 128
earthworms 41
Eciton burchelli (army ant) 101, *102*
ecotypes 7
Ectopistes migratorius (passenger pigeon)
    128, *129*
edge effects
    temperate woodlands 165
    tropical forests 96–7, 107–8, 165
elephants 122
    see also African elephant
emergent trees 94, *95*
emu 16
endangered species
    monitoring 135–53
    spatial distribution of 135–7
endemic species **128**, 137–9
epiphytes 72, 81, 85, 93, 94–5
Erica tetralix (cross-leaved heath) 29, 127
Eucalyptus 129
Euoticus elegantulus (bushbaby) 88
Eurasia, grasslands of 15
European bison (*Bison bonasus*) 163, *166*
extinction
    categories of 152
    see also population extinction; species
    extinction

F

false oat grass (*Arrhenatherum elatius*) 31, 32
fauna, of tropical forests 82, 84
Felis pardalis (ocelot) 99–100
fen 30–31
fen orchid (*Liparis loeselii*) 58, *59*
fertilizers 46–7
Festuca spp. 14, 15, 29
Festuca rubra 32
Filipendula spp. (meadow sweet) 33
fire, grassland 12, 13, *155*
Flectonotus fitzgeraldi (tree frog) *87*
flexi-netting 56, 57
floodplain grasslands *see* wet grasslands
flora, of tropical forests 77–82
food chains 101–2
food partitioning 87–8
food webs 101, 103, *104*
forbs 5
Formicariidae (antbirds) 101, *102*
fossils, tropical forest plants 74, 75
fragmentation of habitat

grasslands 43
    tropical forests 95, 106–8
frogs, of tropical forests *87*, 88–9
fundamental niche 110

G

Galago spp. (bushbaby) 88
Galapagos Islands 128
Gallinago gallinago (snipe) 58, *59*
Game Conservancy Trust 45, 51
gamma diversity 79
gap analysis 110–11
gap effects (tropical forests) 96
genetic diversity 109
genets, defined 141
Gentianella anglica (early gentian) 128
Gentianella germanica (Chiltern gentian) 54
geographical isolation of habitats 139–40
geographical range of species **136**–7
giant anteater (*Myrmecophaga tridactyla*) 6
giant panda (*Ailuropoda melanoleuca*) 135, 146
Gilbertiodendron dewevrei (limbali) 80
global warming 47
grasses, perennial 11
grasshoppers 14
grasslands 5
    biodiversity 6, 7
    economic activities in 6
    factors affecting distribution of 10–13
    as global resource 7–9
    global variation in 14–16
    importance of 5–7
    management 31, 52–3, 133–4, 156–62
    need for remedial action 49–50
    wildlife habitat 6, 7
    see also British grasslands
grassland ecosystems
    pressures on 42–7
        effects of agriculture on biodiversity 44–7
        effects of hunting on biodiversity 43–4
        fragmentation 43
    structure and dynamics of 35–42
        primary production in 36
        role of detritivores in 41
        role of herbivores in 36–40
grazing 18–19
    of calcareous grasslands 55–6, 57
    effect on grassland distribution 12, 13
grazing animals *see* herbivores
grazing ecosystems 35
great crested newt (*Triturus cristatus*) 51
greater horseshoe bat (*Rhinolophus
    ferrumequinum*) 41
greater plantain (*Plantago major*) 33

green-winged orchid (*Orchis morio*) 136, *137*, 140–42, 151
grey partridge (*Perdix perdix*) 45, 46
grey wolf (*Canis lupus*) 14, 129–30
Gryllotalpa gryllotalpa (mole cricket) 58

H

habitat 122, 123–4
    geographical isolation of 139–40
habitat loss 124
    British temperate woodland 124–8
habitat management 133–4, 155–67
Hamearis lucina (Duke of Burgundy butterfly) 57
harvesting, species loss from 131
hawthorn (*Crataegus monogyna*) 159
hay meadows 19, 20
hazel (*Corylus avellana*) 162
heather moorlands 29–30
heathlands 127, 128, 157
hedgehog 44
hedgerows 45
herbicides 45
herbivores 98
    effect on nutrient cycling 37, 38–9
    insect pollinators 40
    selective grazing by 36–7
    trampling by 38–9
Hesperia comma (silver-spotted skipper) 156
heterotrophs 98
high brown fritillary (*Argynnis adippe*) 38, 162
Hippocrepis comosa (horseshoe vetch) 57, 127
hogplum tree (*Spondias mombin*) 109
hogweed 32
Holcus lanata (Yorkshire fog) 32
Holkham National Nature Reserve 61–2
horseshoe vetch (*Hippocrepis comosa*) 57, 127
hound's tongue (*Cynoglossum officinale*) 159
human activities
    effect on grassland distribution 12, 13, 18–20
    effect on tropical forests 106
hummingbirds 87
hunting
    effects on biodiversity 43–4
    species loss from 128–30
Hymenaea courbaril 94

I

insects
    pollinators 40
    of tropical forests 84, 87, 103, *104*
interspecific competition 84
introduced species 44

invertebrates
    and nutrient cycling 40
    pollinators 40
island species 128
isolation *see* geographical isolation of habitats
Ithominae (antbutterflies) 101–2

J

jaguar (*Panthera onca*) 15, 99
Juncus spp. (rushes) 29, 33
juniper (*Juniperus* spp.) 159, 160

K

kangaroos 16
koala (*Phascolarctos cinereus*) 128–9

L

lady's slipper orchid (*Cypripedium* sp.) 135
Lagopus lagopus scotica (red grouse) 30
land crabs 92
Landsat satellites 75–6
lapwing (*Vanellus vanellus*) 47, 58, *59*
large blue butterfly (*Maculinea arion*) 42–3, 54
leaf miners 103, *104*
leaf-litter, in tropical forests 92, 105
Lepidoptera, temporal resource partitioning in 85
lesser whitethroat (*Sylvia curruca*) 57
lianas 80
Limax cineroniger (slug) *126*
Limax tenellus (slug) *126*
limbali (*Gilbertiodendron dewevrei*) 80
Liparis loeselii (fen orchid) 58, *59*
logging, selective 109–10, *111*
Lolium perenne (perennial rye-grass) 31, 46
long-tailed marmot (*Marmota caudata*) 15
lorises 88
Loxodonta africana (African elephant), estimation of numbers 146–7
Luscinia megarhynchos (nightingale) 57
Lysandra bellargus (adonis blue butterfly) 56

M

Maculinea arion (large blue butterfly) 42–3, 54
mammals
    global hotspots for *138*
    of tropical forests 85, *86*, 95
management of habitats/species 133–4
    of grasslands 31, 52–3, 133–4, 156–62
    of sustainable forests 109–10
mark–release–recapture sampling method **147**–9
marmot (*Marmota* spp.) 15

Marmota caudata (red or long-tailed marmot) 15
Martin Down National Nature Reserve 56–7
mat-grass (*Nardus stricta*) 28, 29, *30*
meadows 19, 20, *31*
    wild flower *49*
meadow sweet (*Filipendula* spp.) 33
Mentha pulegium (penny royal) 58
mesotrophic grassland 21–2
    effects of management methods 31
    microhabitats in 31–3
microhabitats, in mesotrophic grasslands 31–3
Micrurius dissoleucus (coral snake) *94*
military orchid (*Orchis militaris*) 54
military training areas *50*
mimicry 102–103
minimum viable population size (MVP) 151–2
mole cricket (*Gryllotalpa gryllotalpa*) 58
monitoring (biological) 131–2
    of endangered species 135–53
Monks Wood National Nature Reserve 164–5
monodominant trees 79–80
montane grassland 21, 24–6
mora (*Mora excelsa*) 79
Müllerian mimicry 102–3
muntjac deer (*Muntiacus muntjak*) 93
mycorrhizas 105
Myrmecophaga tridactyla (giant anteater) 6
Myrmica sabuleti (red ant) 42–3, 54
myxomatosis 20, 56, 158

N

Nardus stricta (mat-grass) 28, 29, *30*
National Vegetation Classification (NVC) 21–6, 28, 51
natural grasslands 7–9
Nepenthes spp. (pitcher plants) 87
nettle (*Urtica dioica*) 32
niche 28
nightingale (*Luscinia megarhynchos*) 57
nitrogen cycling 37, 38–9
nocturnal species **85**
nomenclature, biological 122
North America, grasslands of 14
nutrient cycling 37, 38–9, 40
    in tropical forests 104–5
NVC *see* National Vegetation Classification

O

oaks 122, *123*
ocelot (*Felis pardalis*) 99–100
Oedemera nobilis (pollen beetle) *40*
orchids 136

see also fen orchid; green-winged orchid; lady's slipper orchid; military orchid
Orchis militaris (military orchid) 54
Orchis morio (green-winged orchid) 136, *137*, 140–42, 151
owl butterfly (*Caligo* sp.) 93, *94*, 102
oxlip (*Primula elatior*) 126
Ozotoceros bezoarticus (pampas deer) 15

P

paca (*Agouti paca*) 93, *94*
Pachysima genus (ants) 80
pampas 8, *9*, 14
pampas deer (*Ozotoceros bezoarticus*) 15
Panatal, Brazil, wetlands 11
Panthera onca (jaguar) 15, 99
parasitoids 103, *104*
passenger pigeon (*Ectopistes migratorius*) 128, 129
pasture 19, 20, *31*, 33
pear, wild (*Dumboya* sp.) *8*
pearl-bordered fritillary (*Boloria euphrosyne*) 162
pedunculate oak (*Quercus robur*) 122, *123*
Peltogyne pubescens 94
Pennisetum spp. 15
penny royal (*Mentha pulegium*) 58
Pentaclethra macroloba 95
Perdix perdix (grey partridge) 45, 46
perennial rye-grass (*Lolium perenne*) 31, 46
Periodic Block system 109
Peripatus (velvet worm/walking worm) 92
Perodictus potto (potto) 88
pesticides 45
Phascolarctos cinereus (koala) 128–9
phosphorus, limiting nutrient 29
Phragmites australis (common reed) 30
Phrynus longipes (whip scorpion) 100
Pilch Field, green-winged orchids in 140–42, 151
pioneer species **96**
pitcher plants (*Nepenthes* spp.) 87
plagioclimax *see* deflected climax vegetation
Plantago lanceolata (ribwort plantain) 33
Plantago major (greater plantain) 33
plants, global hotspots for *138*, 139
Plectrachne spp. 16
Poa pratensis (smooth meadow-grass) 33
Poa trivialis (rough meadow-grass) 32
poaching of animals 44
poaching of soil **38**
poisonous arrow tree frog (*Dendrobates truncatus*) *94*
pollen analysis 131

pollen beetle (*Oedemera nobilis*) *40*
pollination 40
Polygala calcarea (chalk milkwort) 54
populations 122
    counting individuals of 142–3
    sampling 143–9
    size of
        monitoring 140–42
        using data on 150–53
population extinction 124
    see also probability of extinction
population viability analysis (PVA) 151–2
potto (*Perodictus potto*) 88
prairie 8, 9, 13, 14
primary (virgin) forests **71**–2, 163–4, 165–6
primary producers 98
primary succession 155
primates, of tropical forests 95, *96*, 130
Primula elatior (oxlip) 126
Primula veris (cowslip) 57
probability of extinction 150–51, 153
pronghorn (*Antilocapra americana*) 14
protected species/habitats 132–3
pseudo-copulation 136
pyramid of biomass 99

Q

quaking grass (*Briza media*) 28
Quercus petraea (sessile oak) *123*
Quercus robur (pedunculate oak) 122, *123*

R

rabbits 20, 156–60
radar mapping 75–6
radio-telemetry 99
ragwort (*Senecio jacobaea*) 159
rainforests 72–3
red ant (*Myrmica sabuleti*) 42–3, 54
Red Data Book, British dung beetles 41
red deer 163, *166*
red grouse (*Lagopus lagopus scotica*) 30
red howler monkey (*Alouatta seniculus*) 95, *96*
Red List of Threatened Species 152–3, *154*
red marmot (*Marmota caudata*) *15*
redshank (*Tringa totanus*) 58, *59*
reed, common (*Phragmites australis*) 30
reforestation 108
refugia 89, 90–91
reptiles, global hotspots for *139*
resource partitioning 59, 84–5
    breeding sites 88–9
    food 87–8
    spatial 85–7, 88, 89
    temporal 85, *86*, 89

rhea (*Rhea americana*) 15
rhinoceros see white rhinoceros
Rhinolophus ferrumequinum (greater horseshoe bat) 41
ribwort plantain (*Plantago lanceolata*) 33
rotational grazing 56
rough meadow-grass (*Poa trivialis*) 32
Royal Society for the Protection of Birds (RSPB) 45, 51
Rubus spp. (brambles) 33
Rumex spp. (docks) 32
rushes (*Juncus* spp.) 29, 33
rye-grass, perennial (*Lolium perenne*) 31, 46

S

Sahel 8, 15, 47
Salix spp. (willows) *30*, 31
sampling of populations 143–9
    by direct counting 143–7
    by using traps 147–9
satellite imaging 75–6
Satyridae 93, 95
savanna 8, 10, 13, *155*
Schoenefeldia spp. 15
scrub clearance 55, 57
secondary forests **71**, 72
secondary succession 155, 156, 157
sedimentary analysis 131
semi-natural grasslands 12, 18, 21
Senecio jacobaea (ragwort) 159
sessile oak (*Quercus petraea*) *123*
sheep 19, 20, 55–6
shrubby seablite (*Suaeda vera*) 156
silver-spotted skipper (*Hesperia comma*) 156
Simpson diversity index 160–62
Skokholm 157, 158
skylark (*Alauda arvensis*) 45, 46
slubbing 60
slugs 126
small-leaved lime (*Tilia cordata*) 125, 126
smooth meadow-grass (*Poa pratensis*) 33
snipe (*Gallinago gallinago*) 58, *59*
soil communities, in tropical forests 92, *104*
soils, and grassland communities 29
Sorghum spp. 16
South America, grasslands of 14–15
spatial distribution of species 135–7
spatial partitioning of resources 85–7, 88, 89
    see also stratification in tropical forests
species 122
species coexistence 84–9
species diversity 50, 77, 109, 160
    in tropical forests 89–91
    see also Simpson diversity index

species extinction 7, **124**, 128

species richness 50, 77, 78–9, 82, 84, 109, 164
in tropical forests 89–91

species vulnerability, measure of 152–3

spinifex grasses 16

Spondias mombin (hogplum tree) 109

Spruce, Richard 73

steppe 8, *9*, 11, 14, 15

Stipa spp. 14, 15

stochastic processes **106**

storage ecosystems 35

strangler plants 80–81

stratification in tropical forests 91–7

Suaeda fruticosa see Suaeda vera

Suaeda vera (shrubby seablite) 156

subspecies 122

succession 72
control of 133–4, 155

Sus scrofa (wild boar) 163, *166*

Sussex, changes in landscape 127

sustainable development 109

sustainable forest management 109–10

swallowtail butterflies, species richness and climate 90

Swartzia sp. *82*

Sylvia curruca (lesser whitethroat) 57

T

tamarind tree (*Tamarindus indicus*) 78

Taraxacum spp. (dandelion) 32

Taxus spp. (yews) 159

teak (*Tectona grandis*) 108

temperate grasslands 8–9, 10–11

temperate woodland
loss in Britain 124–8
see also Monks Wood National Nature Reserve

temporal partitioning of resources 85, *86*, 89

Thesium humifusum (bastard toadflax) 54

thistles 32

threatened species, categories of 152–3, *154*

thyme, wild (*Thymus polytrichus*) 42, *43*

tiger butterfly (*Tithorea harmonia*) 102

Tilia cordata (small-leaved lime) *125*, 126

Tithorea harmonia (tiger butterfly) 102

toads
of tropical forests 88–9
cane toad (*Bufo marina*) 44

top predators 99–100, *101*

Total Station System (surveying equipment) 140

trampling of grasslands 38–9, 55

traps, sampling using 147–9

tree frogs
Dendrobates tinctorius 94
Flectonotus fitzgeraldi 87

trees
emergent 94, *95*
of tropical forests 72, 78–82

Tringa totanus (redshank) 58, *59*

Triodia spp. 16

Triturus cristatus (great crested newt) 51

tropical forests 71–3, 130–31
biodiversity of 77–84
future of 106–11
global extent of 74–5
layers in (stratification) 91–7
canopy 94–5, *96*
gaps and edges 96–7, 107–8, 165
soils and leaf-litter 92, 105
understorey 82, 93, *94*
origins of species richness and diversity in 89–91
trophic levels in 98–100

tropical grasslands 8, 10, 11

tussock grasslands 11, 16, 32

U

UK Biodiversity Action Plan 51

understorey layer in tropical forests **82**, 93, *94*

undisturbed forests *see* primary forests

Ursus arctos (brown bear) 129–30

Urtica dioica (nettle) 32

V

Vanellus vanellus (lapwing) 47, 58, *59*

veld 8, *9*

velvet worm (*Peripatus*) 92

Viola riviniana (dog violet) 38

virgin forests *see* primary forests; wildwood

W

walk-and-count sampling system 143–4

walking worm (*Peripatus*) 92

wallabies 16

Wallace, Alfred Russel 77–8, 82, 83, 103

Wallace's Line 83

warning coloration 93, *94*, 102

waterlogging of grasslands 11

wet fences 62

wet grasslands 30–31, 47, 58–62
water-level control in 60

wetlands 11

wheat 9

whip scorpion (*Phrynus longipes*) 100

white-faced capuchin monkey (*Cebus capucinus*) 110, *111*

white rhinoceros (*Ceratotherium simum*) 16

wild boar (*Sus scrofa*) 163, *166*

wild pear (*Dumboya* sp.) *8*

wild thyme (*Thymus polytrichus*) 42, *43*

wildwood
destruction of 125, 126, 131
surviving fragments of 163

willows (*Salix* spp.) *30*, 31

wolf *see* grey wolf

wood anemone (*Anemone nemorosa*) 126

woodland, managed and natural 162–7

Y

yew (*Taxus* spp.) 159

Yorkshire fog (*Holcus lanata*) 32

Z

Zimbabwe, droughts 10

Zofinsky Prales, Czech Republic 133